REVISE PEARSON EDEXCEL INT GCSE (9–1) Mathematics A Higher Tier

REVISION GUIDE

Series Consultant: Harry Smith

Author: Harry Smith

A note from the publisher

In order to ensure that this resource offers high-quality support for the associated Pearson qualification, it has been through a review process by the awarding body. This process confirms that this resource fully covers the teaching and learning content of the specification or part of a specification at which it is aimed. It also confirms that it demonstrates an appropriate balance between the development of subject skills, knowledge and understanding, in addition to preparation for assessment.

Endorsement does not cover any guidance on assessment activities or processes (e.g. practice questions or advice on how to answer assessment questions) included in the resource nor does it prescribe any particular approach to the teaching or delivery of a related course.

While the publishers have made every attempt to ensure that advice on the qualification and its assessment is accurate, the official specification and associated assessment guidance materials are the only authoritative source of information and should always be referred to for definitive guidance.

Pearson examiners have not contributed to any sections in this resource relevant to examination papers for which they have responsibility.

Examiners will not use endorsed resources as a source of material for any assessment set by Pearson.

Endorsement of a resource does not mean that the resource is required to achieve this Pearson qualification, nor does it mean that it is the only suitable material available to support the qualification, and any resource lists produced by the awarding body shall include this and other appropriate resources.

Aiming Higher

This icon helps you spot tricky questions. If you're aiming for a grade 7, 8 or 9 you should make sure you are confident with these questions.

Worked solution video

Worked solution videos

Some of the questions in this Revision Guide have worked solution videos. Use your phone, tablet or webcam to scan the QR code to watch the video.

For the full range of Pearson Edexcel International GCSE and International AS/A Level titles, visit:
www.pearsonglobalschools.com

ALWAYS LEARNING

PEARSON

Contents

. . . .

A small bit of small print

Pearson Edexcel publishes Sample Assessment Material and the Specification on its website. This is the official content and this book should be used in conjunction with it. The questions in *Now try this* have been written to help you practise every topic in the book. Remember: the real exam questions may not look like this.

Factors and primes

The **factors** of a number are any numbers that divide into it exactly. A **prime number** has exactly two factors. The prime numbers are 2, 3, 5, 7, 11, 13, 17, 19 and so on.

Prime factors

If a number is a factor of another number **and** it is a prime number then it is called a **prime factor**. You use a factor tree to find prime factors.

Remember to circle the prime factors as you go along. The order doesn't matter.

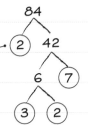

$$84 = 2 \times 2 \times 3 \times 7$$ — Remember to put in the multiplication signs.
$$= 2^2 \times 3 \times 7$$ — This is called a **product** of **prime factors**.

The highest common factor (HCF) of two numbers is the **highest number** that is a **factor** of both numbers.

The lowest common multiple (LCM) of two numbers is the **lowest number** that is a **multiple** of both numbers.

Worked example

(a) Write 108 as the product of its prime factors. Give your answer in index form. **(3 marks)**

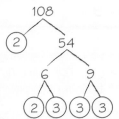

$$108 = 2 \times 2 \times 3 \times 3 \times 3 = 2^2 \times 3^3$$

(b) Work out the highest common factor (HCF) of 108 and 24. **(2 marks)**

$$108 = ②\times②\times③\times 3 \times 3$$
$$24 = ②\times②\times 2 \times ③$$
HCF is $2 \times 2 \times 3 = 12$

(c) Work out the lowest common multiple (LCM) of 108 and 24. **(2 marks)**

LCM = $12 \times 3 \times 3 \times 2 = 216$

Draw a factor tree. Continue until every branch ends with a prime number. This question asks you to write your answer in **index form**. This means you need to use **powers** to say how many times each prime number occurs in the product.

Check it!
$2^2 \times 3^3 = 4 \times 27 = 108$ ✓

To find the HCF, circle all the prime numbers which are **common** to both products of prime factors. 2 appears twice in both products so you have to circle it twice. Multiply the circled numbers together to find the HCF.

To find the LCM multiply the HCF by each unshared prime from part (b).

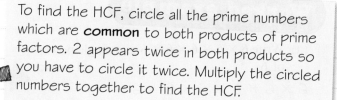

You can use numbers given in index form directly:
HCF: Choose the **lowest** power of each prime
LCM: Choose the **highest** power of each prime

Now try this

1 (a) Express 980 as a product of its prime factors. **(3 marks)**

  (b) Find the highest common factor (HCF) of 980 and 56. **(2 marks)**

2 $X = 2 \times 3^5 \times 7^2$ $Y = 3^2 \times 5 \times 7$

  (a) Find the highest common factor (HCF) of X and Y. **(2 marks)**

  (b) Find the lowest common multiple (LCM) of X and Y. **(2 marks)**

Indices 1

The index laws tell you how to work with **powers** of numbers.

1 Index laws

Indices include square roots, cube roots and powers.

You can use the index laws to simplify powers and roots.

$$a^m \times a^n = a^{m+n}$$
$$4^3 \times 4^7 = 4^{3+7} = 4^{10}$$

$$\frac{a^m}{a^n} = a^{m-n}$$
$$12^8 \div 12^3 = 12^{8-3} = 12^5$$

$$(a^m)^n = a^{mn}$$
$$(7^3)^5 = 7^{3 \times 5} = 7^{15}$$

2 Cube root

The cube root of a positive number is positive.

$$4 \times 4 \times 4 = 64$$
$$4^3 = 64$$
$$\sqrt[3]{64} = 4$$

The cube root of a negative number is negative.

$$-4 \times -4 \times -4 = -64$$
$$(-4)^3 = -64$$
$$\sqrt[3]{-64} = -4$$

3 Powers of 0 and 1

Anything raised to the power 0 is equal to 1.

$$6^0 = 1 \quad 1^0 = 1 \quad 7223^0 = 1 \quad (-5)^0 = 1$$

Anything raised to the power 1 is equal to itself.

$$8^1 = 8 \quad 499^1 = 499 \quad (-3)^1 = -3$$

Indices checklist

- ☑ The base numbers have to be the same.
- ☑ If there's no index, the number has the power 1.
- ☑ Be careful with negatives: $(-3)^2 = 9$

Worked example

(a) Write $6 \times 6 \times 6 \times 6 \times 6$ as a single power of 6. **(1 mark)**

$$6 \times 6 \times 6 \times 6 \times 6 = 6^5$$

(b) Simplify $\frac{3^8 \times 3}{3^4}$ fully, leaving your answer in index form. **(2 marks)**

$$\frac{3^8 \times 3}{3^4} = \frac{3^9}{3^4} = 3^5$$

3 is the same as 3^1. For part (b), use the rule $a^m \times a^n = a^{m+n}$ to simplify the numerator, then use $\frac{a^m}{a^n} = a^{m-n}$ to simplify the fraction. Remember to write down both steps of your working and give your answer as a power.

Learn it!

You need to be able to recognise the square numbers up to 15^2 and the cubes of 1, 2, 3, 4, 5 and 10. You can check with a calculator but you'll be more confident if you learn them.

For 1(b), start by working out $\frac{9625}{7 \times 11}$

Now try this

1 (a) Write $7^3 \times 7^5$ as a single power of 7. **(1 mark)**

   (b) $9625 = 5^n \times 7 \times 11$
   Find the value of n. **(2 marks)**

2 $(\sqrt[3]{-27})^k = 9$
   Write down the value of k. **(2 marks)**

3 (a) Simplify, leaving your answers in index form
   (i) $\frac{2^9}{2^5}$ (ii) $(7^2)^6$ (iii) $5^2 \times 5^0$ **(3 marks)**

   (b) $\frac{3^n}{3^2 \times 3^5} = 3^4$
   Find the value of n. **(2 marks)**

Indices 2

You can use these index laws to deal with powers that are **fractions** or **negative numbers**.

 Negative powers

$$a^{-n} = \frac{1}{a^n}$$

$$5^{-2} = \frac{1}{5^2} = \frac{1}{25}$$

Be careful!

A **negative** power can still have a **positive** answer.

 Reciprocals

$$a^{-1} = \frac{1}{a}$$

This means that a^{-1} is the **reciprocal** of a.
You can find the reciprocal of a fraction by turning it upside down.

$$\left(\frac{5}{9}\right)^{-1} = \frac{9}{5}$$

 Powers of fractions

$$\left(\frac{a}{b}\right)^n = \frac{a^n}{b^n}$$

$$\left(\frac{3}{10}\right)^2 = \frac{3^2}{10^2} = \frac{9}{100}$$

 Combining rules

You can apply the rules one at a time.

$$\left(\frac{a}{b}\right)^{-n} = \left(\frac{b}{a}\right)^n = \frac{b^n}{a^n}$$

$$\left(\frac{2}{3}\right)^{-3} = \left(\frac{3}{2}\right)^3 = \frac{3^3}{2^3} = \frac{27}{8}$$

 Fractional powers

You can use fractional powers to represent roots.

$$a^{\frac{1}{2}} = \sqrt{a} \qquad 49^{\frac{1}{2}} = 7$$
$$a^{\frac{1}{3}} = \sqrt[3]{a} \qquad 27^{\frac{1}{3}} = 3$$
$$a^{\frac{1}{4}} = \sqrt[4]{a} \qquad 16^{\frac{1}{4}} = 2$$

Check it!

A whole number raised to a power less than 1 gets smaller.

 More complicated indices

You can use the index laws to work out more complicated fractional powers.

$$a^{\frac{m}{n}} = \left(a^{\frac{1}{n}}\right)^m$$

Do these calculations **one step at a time.**

$$27^{-\frac{2}{3}} = (27^{\frac{1}{3}})^{-2}$$
$$= (\sqrt[3]{27})^{-2}$$
$$= 3^{-2} = \frac{1}{3^2} = \frac{1}{9}$$

Worked example   Aiming higher

Find the value of n when $3^n = 9^{-\frac{3}{2}}$
Show each step of your working clearly.   **(3 marks)**

$$9^{-\frac{3}{2}} = (3^2)^{-\frac{3}{2}}$$

$$= 3^{2 \times -\frac{3}{2}} = 3^{-3}$$

So $3^n = 3^{-3}$ and $n = -3$

Problem solved!

3^n is not the same as $3n$. You can't divide by 3 to get n on its own. You need to make the base on the right-hand side the same as the base on the left-hand side.
1. Write 9 as a power of 3. Remember to use brackets.
2. Use $(a^n)^m = a^{nm}$ to write the right-hand side as a single power of 3.
3. Compare both sides and write down the value of n.

> You will need to use problem-solving skills throughout your exam – **be prepared!**

Now try this

 Aiming higher

1 You are given that $x = 7^h$ and $y = 7^k$
Write each of the following as a single power of 7.

(a) $\dfrac{x}{y}$ **(1 mark)**

(b) x^2 **(1 mark)**

(c) xy^2 **(2 marks)**

2 Given that $81^{-\frac{3}{4}} = 3^n$, find the value of n.
    (3 marks)

3 Write $\sqrt{\dfrac{49}{7^3}}$ as a single power of 7.
Show every step of your working clearly.   **(3 marks)**

Start by writing 49 as a power of 7

Calculator skills 1

These calculator keys are really useful.

| x^2 | Square a number. | $(-)$ | Enter a negative number. |

| x^3 | Cube a number. | $\sqrt{\Box}$ | Find the square root of a number. |

| x^{-1} | Find the reciprocal of a number. | $\sqrt[3]{\Box}$ | Find the cube root of a number. You might need to press the shift key first. |

| Ans | Use your previous answer in a calculation. | S⇔D | Change the answer from a fraction or surd to a decimal. Not all calculators have this key. |

Rounding rules

1 To **round** a number, you look at the next digit on the right.

5 or more → round up   less than 5 → round down

2 Decimals can be rounded to a given number of **decimal places** (d.p.).

6.475 = 6.48 correct to 2 d.p.

3 To write a number correct to **3 significant figures** (3 s.f.), look at the fourth significant figure.

0.003 07$\underline{9}$ = 0.003 08 to 3 s.f.

4 Zeroes at the start of a decimal number are not counted as significant.

5 Remember that the rule for significant figures still applies to **whole numbers**.

27 = 30 to 1 s.f.

Worked example

You won't be asked questions like the ones on this page in your exam, but you need to make sure you are confident using your calculator.

(a) Work out the value of $\dfrac{\sqrt{8.3}}{12.5 - 7.3}$

Give your answer as a decimal.
Write down all the figures on your calculator display. **(2 marks)**

$$\frac{\sqrt{8.3}}{12.5 - 7.3} = \frac{2.880\,97\ldots}{5.2} = 0.554\,033\,088$$

(b) Give your answer to part (a) correct to 2 significant figures. **(1 mark)**

0.55 (2 s.f.)

Calculate $\sqrt{8.3}$ using the $\sqrt{\Box}$ key. Always show what the top of the fraction comes to as well as the bottom. Remember to write down **all** the figures on your calculator display.

Check it!

Do the whole calculation in one go on your calculator using the ▦ key.

Now try this

(a) Work out the value of $\dfrac{6.1 + 7.5}{1.8^2}$

Give your answer as a decimal.

Write down all the figures on your calculator display. **(2 marks)**

(b) Give your answer to part (a) correct to 3 significant figures. **(1 mark)**

Make sure you write down three significant figures even if the last digit is a zero.

Fractions

You need to be able to work with fractions and mixed numbers confidently **without a calculator**.

① Adding or subtracting fractions

Convert mixed numbers to improper fractions

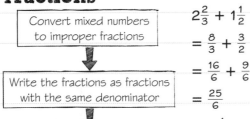

$2\frac{2}{3} + 1\frac{1}{2}$

$= \frac{8}{3} + \frac{3}{2}$

Write the fractions as fractions with the same denominator

$= \frac{16}{6} + \frac{9}{6}$

$= \frac{25}{6}$

Add or subtract the fractions

$= 4\frac{1}{6}$

If you have an improper fraction then convert to a mixed number

③ Dividing fractions

Convert any mixed numbers to improper fractions

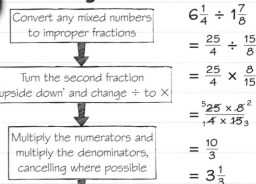

$6\frac{1}{4} \div 1\frac{7}{8}$

$= \frac{25}{4} \div \frac{15}{8}$

Turn the second fraction 'upside down' and change ÷ to ×

$= \frac{25}{4} \times \frac{8}{15}$

$= \frac{\overset{5}{\cancel{25}} \times \overset{2}{\cancel{8}}}{\underset{1}{\cancel{4}} \times \underset{3}{\cancel{15}}}$

Multiply the numerators and multiply the denominators, cancelling where possible

$= \frac{10}{3}$

$= 3\frac{1}{3}$

Convert any improper fractions to mixed numbers

② Multiplying fractions

Convert any mixed numbers to improper fractions

$3\frac{1}{4} \times 2\frac{2}{3}$

$= \frac{13}{4} \times \frac{8}{3}$

Multiply the numerators and multiply the denominators, cancelling where possible

$= \frac{13 \times \overset{2}{\cancel{8}}}{\underset{1}{\cancel{4}} \times 3}$

$= \frac{26}{3}$

Convert any improper fractions to mixed numbers

$= 8\frac{2}{3}$

Worked example

Show that $7\frac{1}{3} - 2\frac{3}{4} = 4\frac{7}{12}$ **(3 marks)**

$7\frac{1}{3} - 2\frac{3}{4} = \frac{22}{3} - \frac{11}{4}$

$= \frac{88}{12} - \frac{33}{12}$

$= \frac{55}{12}$

$= 4\frac{7}{12}$

Worked example

The diagram shows three identical shapes. $\frac{3}{4}$ of shape **A** is shaded and $\frac{3}{5}$ of shape **C** is shaded.

A   **B**   **C**

What fraction of shape **B** is shaded?   **(3 marks)**

$1 - \frac{1}{4} - \frac{2}{5} = \frac{20}{20} - \frac{5}{20} - \frac{8}{20} = \frac{7}{20}$

Show that...

In your International GCSE exam, a calculation involving fractions or mixed numbers might say "Show that...". This means you have to:

- ✓ Start by writing down the left-hand-side of the calculation
- ✓ Show every step of your working clearly
- ✓ Finish by writing down the answer.

Now try this

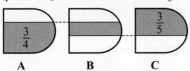

1 Show that

(a) $\frac{7}{10} - \frac{1}{4} = \frac{9}{20}$ **(2 marks)**

(b) $3\frac{4}{9} + 1\frac{5}{6} = 5\frac{5}{18}$ **(3 marks)**

(c) $\frac{3}{4} \div \frac{5}{12} = 1\frac{4}{5}$ **(2 marks)**

(d) $1\frac{7}{8} \times 2\frac{2}{3} = 5$ **(3 marks)**

2 Three girls shared a full bottle of cola.
Ling drank $\frac{1}{4}$ of the bottle.
Dunia drank $\frac{3}{10}$ of the bottle.
Megan drank the rest.

(a) Work out the fraction of the bottle of cola that Megan drank.   **(3 marks)**

Dunia drank 36 cl of cola.

(b) How much cola was in the full bottle?   **(2 marks)**

Worked solution video

Decimals

Terminating decimals can be written exactly. You can write a terminating decimal as a fraction with denominator 10, 100, 1000, and so on.

$$0.24 = \frac{24}{100} = \frac{6}{25}$$

Recurring decimals have one digit or group of digits repeated forever. You can use dots to show the recurring digit or group of digits.

$$\frac{2}{3} = 0.6666... = 0.\dot{6}$$

The dot tells you that the 6 repeats forever.

$$\frac{346}{555} = 0.623\,4234... = 0.6\dot{2}3\dot{4}$$

These dots tell you that the group of digits 234 repeats for ever.

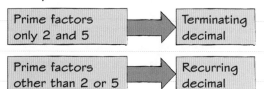

Recurring or terminating?

To check whether a fraction produces a recurring decimal or a terminating decimal, write it in its simplest form and find the prime factors of its **denominator**.

| Prime factors only 2 and 5 | → | Terminating decimal |
| Prime factors other than 2 or 5 | → | Recurring decimal |

You could also write the denominator as a product of prime factors:

$$\frac{7}{50} = \frac{7}{2 \times 5^2}$$

The only factors are 2 and 5 so $\frac{7}{50}$ produces a terminating decimal.

Worked example

(a) Show that $\frac{7}{50}$ can be written as a terminating decimal. **(1 mark)**

$$\frac{7}{50} = \frac{14}{100} = 0.14$$

(b) Show that $\frac{11}{24}$ **cannot** be written as a terminating decimal. **(2 marks)**

$$\frac{11}{24} = \frac{11}{2^3 \times 3}$$

Denominator contains a factor other than 2 or 5 so decimal is recurring.

Worked example

(a) Show that $\frac{2}{9}$ is equivalent to 0.222… **(1 mark)**

$$\begin{array}{r} 0.2\ 2\ 2... \\ 9\overline{)2.\,{}^20{}^20{}^20...} \end{array}$$

(b) Hence, or otherwise, show that $0.7\dot{2} = \frac{13}{18}$ **(3 marks)**

$$0.7222... = 0.222... + 0.5$$
$$= \frac{2}{9} + \frac{1}{2}$$
$$= \frac{4}{18} + \frac{9}{18} = \frac{13}{18}$$

You could also use long division for part (a).

Fractions and decimals

To convert a fraction into a decimal, divide the numerator by the denominator.

$$\frac{2}{5} = 2 \div 5 = 0.4$$

It's useful to remember these common fraction-to-decimal conversions:

Fraction	$\frac{1}{100}$	$\frac{1}{20}$	$\frac{1}{10}$	$\frac{1}{2}$	$\frac{1}{5}$	$\frac{1}{4}$	$\frac{3}{4}$
Decimal	0.01	0.05	0.1	0.5	0.2	0.25	0.75

Now try this

1 By writing 250 as a product of its prime factors, explain why $\frac{1}{250}$ can be written as a terminating decimal. **(2 marks)**

2 Show that $\frac{1}{140}$ **cannot** be written as a terminating decimal. **(2 marks)**

3 Show that $\frac{5}{11} = 0.\dot{4}\dot{5}$ **(2 marks)**

Recurring decimals

You can use algebra to convert a recurring decimal into a fraction. Here is the strategy:

Write the recurring decimal as n. ➡ Multiply by 10, 100 or 1000. ➡ Subtract to remove the recurring part. ➡ Divide by 9, 99 or 999 to write as a fraction.

If you need to do this in your exam you must show **all your working**. For a reminder about recurring decimals have a look at page 6.

Worked example

Aiming higher

Prove that the recurring decimal $0.\dot{2}\dot{4}$ has the value $\frac{8}{33}$ **(2 marks)**

Let $n = 0.242\,424\,24...$

$100n = 24.242\,424\,24...$

$-n = -0.242\,424\,24...$

$99n = 24$

$n = \dfrac{24}{99} = \dfrac{8}{33}$

Some calculators will convert recurring decimals into fractions for you. But the question says 'Prove that …' so you must write down all the steps shown here.

1. Write the recurring decimal equal to n, and write out some of its digits.

2. Multiply both sides by 100 as there are 2 recurring digits.

3. Subtract n to remove the recurring part.

4. Divide both sides by 99 to write n as a fraction.

5. Simplify the fraction.

Multiply by...

10 if 1 digit recurs. ✓

100 if 2 digits recur. ✓

1000 if 3 digits recur. ✓

Worked example

Aiming higher

Show that $0.4\dot{7}\dot{3}$ can be written as the fraction $\frac{469}{990}$ **(2 marks)**

Let $n = 0.473\,737\,37...$

$100n = 47.373\,737\,37...$

$-n = -0.473\,737\,37...$

$99n = 46.9$

$n = \dfrac{46.9}{99}$

$n = \dfrac{469}{990}$

Problem solved!

In this recurring decimal, the digit 4 does not recur. Follow the same steps to write n as a fraction. After you divide by 99, multiply the top and bottom of your fraction by 10 to convert the decimal in the numerator into an integer.

You will need to use problem-solving skills throughout your exam – **be prepared!**

Worked solution video

Scan this QR code to watch a video of this question being solved.

Now try this

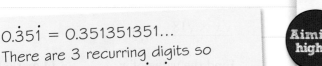

$0.\dot{3}5\dot{1} = 0.351351351...$
There are 3 recurring digits so you need to write $0.\dot{3}5\dot{1}$ as n, then multiply by 1000. You will get a fraction with denominator 999, which you can simplify.

Aiming higher

1 Work out the recurring decimal $0.5\dot{4}$ as a fraction in its simplest form. **(2 marks)**

2 Prove that the recurring decimal $0.0\dot{1}\dot{8}$ has the value $\frac{1}{55}$ **(2 marks)**

3 Show that $0.\dot{3}5\dot{1}$ can be written as the fraction $\frac{13}{37}$ **(2 marks)**

Surds 1

You can give exact answers to calculations by leaving some numbers as square roots.

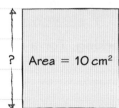

? | Area = 10 cm²

This square has a side length of $\sqrt{10}$ cm.

You can't write $\sqrt{10}$ exactly as a decimal number. It is called a **surd**.

Rules for simplifying square roots

These are the most important rules to remember when dealing with surds.

1 $\sqrt{ab} = \sqrt{a} \times \sqrt{b}$ $\sqrt{8} = \sqrt{4} \times \sqrt{2} = 2\sqrt{2}$

2 $\sqrt{\dfrac{a}{b}} = \dfrac{\sqrt{a}}{\sqrt{b}}$ $\sqrt{\dfrac{3}{25}} = \dfrac{\sqrt{3}}{\sqrt{25}} = \dfrac{\sqrt{3}}{5}$

You need to remember these rules for your exam.

Worked example

Aiming higher

Show that $\sqrt{45} = 3\sqrt{5}$
Show each stage of your working clearly. **(2 marks)**

$\sqrt{45} = \sqrt{9 \times 5}$
$= \sqrt{9} \times \sqrt{5}$
$= 3\sqrt{5}$

This question says 'Show that ...' so you need to show each step of your working clearly:

1. Look for a factor of 45 which is a square number: $45 = 9 \times 5$
2. Use the rule $\sqrt{ab} = \sqrt{a} \times \sqrt{b}$ to split the square root into two square roots.
3. Write $\sqrt{9}$ as a whole number.

Rationalising the denominator of a fraction means making the denominator a whole number.

You can do this by multiplying the top **and** bottom of the fraction by the surd part in the denominator.

$$\frac{5}{3\sqrt{2}} \xrightarrow{\times\sqrt{2}} \frac{5\sqrt{2}}{6}$$

The surd part of the denominator is $\sqrt{2}$

Remember that $\sqrt{2} \times \sqrt{2} = 2$
So $3\sqrt{2} \times \sqrt{2} = 3 \times 2 = 6$

Good form

Most surd questions ask you to write a number or answer in a certain **form**.

This means you need to find **integers** for all the letters in the expression.

$6\sqrt{3}$ is in the form $k\sqrt{3}$
   $k = 6$

The integers can be positive or negative.

$4 - 9\sqrt{2}$ is in the form $p + q\sqrt{2}$
   $p = 4$ and $q = -9$

You can check your answer by writing down the integer value for each letter.

Now try this

Find factors of 32 and 98 which are **square** numbers.

Aiming higher

1 Write $\sqrt{32} + \sqrt{98}$ in the form $p\sqrt{2}$ where p is an integer. Show each stage of your working clearly. **(2 marks)**

2 Show that $\dfrac{35}{\sqrt{7}} = 5\sqrt{7}$ **(2 marks)**

Rationalise the denominator by multiplying top and bottom by $\sqrt{7}$

3 x is an integer such that
$$\frac{\sqrt{x} \times \sqrt{18}}{\sqrt{3}} = 8\sqrt{3}$$
Find the value of x. **(4 marks)**

Calculator skills 2

You need to be able to work out basic percentages quickly using your calculator.

Calculating with fractions

You can enter fractions on your calculator using the  key and the arrows. For example, to work out $\frac{3}{8}$ of 52:

▯ 3 ▼ 8 ▶ × 5 2 =

$$\frac{3}{8} \times 52$$
$$\frac{39}{2}$$

If you want to convert an answer on your calculator display from a fraction to a decimal you can use the S⇔D key.

Quick conversions

You can convert between fractions, decimals and percentages quickly using a calculator:

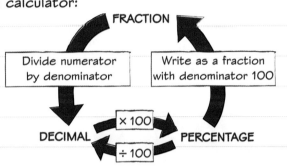

FRACTION

| Divide numerator by denominator | Write as a fraction with denominator 100 |

DECIMAL   × 100   PERCENTAGE
÷ 100

You won't have to answer basic percentages questions like these in your exam, but you might need to use percentage skills in harder questions. To find a percentage of an amount:

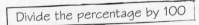

| Divide the percentage by 100 |

↓

| Multiply by the amount |

Worked example

A company gives 3.5% of its profits to charity. In 2017 the company made profits of \$470 000. How much money did the company give to charity in 2017? **(2 marks)**

3.5 ÷ 100 = 0.035

0.035 × 470 000 = 16 450

The company gave \$16 450 to charity.

Worked example

In a year group of 85 students, 62 buy their lunch at school. Express 62 as a percentage of 85. Give your answer correct to 1 decimal place. **(2 marks)**

62 ÷ 85 = 0.72941...

0.72941... × 100 = 72.941...

72.9% of students buy their lunch in school.

To write one quantity as a percentage of another:

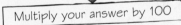

| Divide the first quantity by the second quantity |

↓

| Multiply your answer by 100 |

Always write down at least five digits from your calculator display before rounding your answer.

Now try this

1 Last year, a university had 226 graduates. 195 of them found jobs immediately. Express 195 as a percentage of 226. Give your answer correct to 1 decimal place.
(2 marks)

2 Aisha earns \$2230 per month and spends 25% of it on rent. Javier earns \$1800 per month and spends 30% on rent. Who spends the greater amount on rent?
(3 marks)

You need to show your working clearly to demonstrate your strategy.

Ratio

Ratios are used to compare quantities.

You can find equivalent ratios by multiplying or dividing by the same number.

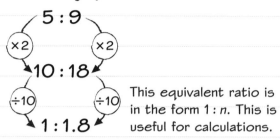

This equivalent ratio is in the form 1 : *n*. This is useful for calculations.

Simplest form

To write a ratio in its simplest form, find an equivalent ratio with the smallest possible whole number values.

Simplest form
5 : 1   10 : 9
2 : 3 : 4

Not simplest form
10 : 2   1 : 0.9
1 : 1.5 : 2

Worked example

Ali and Nabil share a car. They decide to divide the fuel costs in the ratio 3 : 5
Ali pays £78. How much does Nabil pay? **(2 marks)**

$78 \div 3 = 26$

$26 \times 5 = 130$

Nabil pays £130

Work out the value of one part of the ratio first. You could also use equivalent ratios to solve this problem. Work out how many 3s go into 78 to find the multiplier.

3 : 5
×26   ×26
78 : 130

Check it!
£130 + £78 = £208
Divide £208 in the ratio 3 : 5
3 + 5 = 8 parts in the ratio in total
£208 ÷ 8 = £26
£26 × 3 = £78 ✓

Worked example

A university course has 945 applicants (i.e. people applying) for 126 places.

(a) Find the ratio of the number of applicants to the number of places. Give your ratio in the form *n* : 1 **(2 marks)**

$945 \div 126 = 7.5$

Applicants : Places = 945 : 126
= 7.5 : 1

(b) Of the 126 successful applicants, the ratio of males to females is 4 : 5
Work out the number of females on the course. **(2 marks)**

$4 + 5 = 9$
$126 \div 9 = 14$
$14 \times 5 = 70$

There were 70 females on the course.

Work out the total number of parts in the ratio, then divide 126 by this to work out how many people each part represents. Then multiply by 5 to work out the number of females.

Now try this

1 Andre, Becky and Makito share some money in the ratio 3 : 6 : 7
In total Andre and Becky receive £207
Work out the amount of money Makito receives. **(2 marks)**

2 The ratio of Amir's age to Petra's age is 3 : 7
Amir is 9 years old.
(a) Work out Petra's age. **(2 marks)**

The ratio of Karl's age to Yasmina's age is 5 : 2
The sum of their ages is 84.
(b) Work out Yasmina's age. **(2 marks)**

Proportion

Two quantities are in **direct proportion** when both quantities increase at the same rate.

Number of theatre tickets bought   Total cost

3 ×3 9   £135 ×3 £405

Two quantities are in **inverse proportion** when one quantity increases at the same rate as the other quantity decreases.

Average speed   Time taken

40 km/h ×2 80 km/h   2 hours ÷2 1 hour

Divide or multiply?

You can use common sense to work out whether to divide or multiply in proportion questions.

6 people can build a wall in 4 days.

6 × 4 = 24 so 1 person could build the wall in 24 days.

Multiply because it would take 1 person **longer** to build the wall.

24 ÷ 8 = 3 so 8 people could build the wall in 3 days.

Divide because it would take 8 people **less time** to build the wall.

Worked example

Here are the ingredients for apple crumble.

Apple crumble	
Serves 6 people	
900 g apples	90 g butter
180 g sugar	150 g flour

(a) Henry wants to make apple crumble for 11 people.
Work out the amount of sugar he needs.   **(2 marks)**

Amount needed for 1 person = $\frac{180}{6}$ = 30 g

Amount needed for 11 people = 11 × 30 = 330 g

(b) Carla makes an apple crumble using 2250 g of apples. Work out how many people her apple crumble will serve.   **(2 marks)**

2250 ÷ 900 = 2.5
6 × 2.5 = 15
The apple crumble will serve 15 people.

Problem solved!

There are two ways to tackle this question:
• Work out the cost per unit of volume for each option.
• Work out the amount of cola you get for a fixed amount of money for each option.
It doesn't matter which method you use, but you need to show how you arrived at your conclusion. The easiest way to do this is to write down **what you are working out** at each stage.

> You will need to use problem-solving skills throughout your exam – **be prepared!**

Worked example

Aisha wants to buy some cola for a party. She compares the prices of a two-litre bottle and a pack of eight cans.
• Two-litre bottle   £1.29
• Eight 330 ml cans   £1.99
Which option offers the better value?   **(3 marks)**

Cost in pence per ml:
Two-litre bottle: 129 ÷ 2000 = 0.0645
Cans: 199 ÷ (8 × 330) = 0.075 37…
Cola costs less per ml in a two-litre bottle so that is the better value.

Now try this

 Worked solution video

1.25 kg of cheese costs €16.55 in France.
$1\frac{1}{2}$ lb of the same cheese costs £8.97 in England.
In which country is it cheaper to buy the cheese?
Show all of your working.   **(5 marks)**
1 kg = 2.2 lb   £1 = €1.15

> You need to write quantities in the same units before comparing them. Choose lb or kg and choose £ or €. Remember to write a conclusion.

Percentage change

There are two methods that can be used to increase or decrease an amount by a percentage.

Method 1

Work out 26% of $280:

$\frac{26}{100} \times \$280 = \$72.80$

Subtract the decrease:

$\$280 - \$72.80 = \$207.20$

$280

26% OFF

Method 2

Use a multiplier:

$100\% + 30\% = 130\%$

$\frac{130}{100} = 1.3$

So the multiplier for a 30% increase is 1.3:

$400\,g \times 1.3 = 520\,g$

400 g PLUS 30% EXTRA

Worked example

In 2017 a travel company charged $550 for a particular holiday.

In 2018 they increased their prices by 6.2%.

The travel company offers a 10% discount to customers who purchase their holiday online.

Saanvi purchases the holiday online in 2018. Calculate the price that she pays. **(3 marks)**

$550 \times 1.062 = 584.1$
$584.1 \times 0.9 = 525.69$
The discounted price is $525.69

Problem solved!

Read the question carefully – there are two steps in the working:

6.2% increase
The multiplier is $\frac{100 + 6.2}{100} = 1.062$

10% decrease
The multiplier is $\frac{100 - 10}{100} = 0.9$

Make sure you give **units** with your final answer.

> You will need to use problem-solving skills throughout your exam – **be prepared!**

Calculating a percentage increase or decrease

Work out the amount of the increase or decrease

↓

Write this as a percentage of the original amount

Was £60
Now £39

$60 - 39 = 21$

$\frac{21}{60} \times 100 = 35\%$

This is a 35% decrease.

For a reminder about writing one quantity as a percentage of another, have a look at page 9.

A question might ask you to calculate a percentage **profit** or **loss** rather than an increase or decrease.

Now try this

1 A television originally cost 28 000 rupees. In a sale it was reduced in price by 35%. What was the price of the television in the sale? **(3 marks)**

  Reduction means decrease. The multiplier for a 35% decrease would be 0.65.

2 Johan publishes a monthly poetry magazine. In March he printed 1400 copies of his magazine. In April he increased his print run by 15%.

It costs Johan €800 plus 75 cents per copy to print his magazine. He sells each issue for €1.99.

Assuming Johan sells every copy that he prints, calculate his percentage profit in April. **(4 marks)**

Reverse percentages

In some questions you are given an amount **after** a percentage change, and you have to find the **original amount**. To answer questions like this you need to be really confident with **percentage change**. Revise it on page 12.

Using a multiplier

You can use a multiplier to calculate a percentage increase or decrease. If you are given the **final amount** and you need to find the **original amount**, you can **divide by the multiplier**. Here are two examples.

 A sweater is **reduced** in price by 20% in a sale.

Original price £50 × 0.8 Sale price £40 ÷ 0.8

 The average temperature **increases** by 5%.

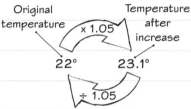

Original temperature 22° × 1.05 Temperature after increase 23.1° ÷ 1.05

Worked example

In a sale, normal prices are reduced by 15%
The sale price of a pair of trainers is $75.65
Work out the normal price of the trainers.

(3 marks)

$100\% - 15\% = 85\%$

$\frac{85}{100} = 0.85$

$75.65 \div 0.85 = 89$

The original price was $89

Read percentages questions carefully so you know what information you have been given. This question tells you the price **after** the percentage decrease, so you need to use **reverse percentages**. To find the multiplier for a 15% decrease:
1. Subtract 15% from 100%
2. Divide by 100 to convert to a multiplier.

You need to **divide** by the multiplier to find the original price.

Check it!
Reduce $89 by 15%: $89 × 0.15 = $13.35
$89 − $13.35 = $75.65 ✓

Problem solved!

Plan your answer before you start. You will need to do **two separate calculations** to find both original heights. Then you need to **compare** the original heights and **write a conclusion**.
To find a multiplier for a percentage increase:
1. add the percentage to 100%
2. divide by 100 to convert to a multiplier.
You need to **divide** by the multiplier to find the height in 2013.

 You will need to use problem-solving skills throughout your exam – **be prepared!**

Worked example

Amy and Elian measure their heights each year. This table shows their heights in 2014.

	Height in 2014 (cm)	Percentage increase since 2013
Elian	154	10%
Amy	150.8	4%

Who was taller in 2013? Give reasons for your answer.   **(4 marks)**

Elian's original height = 154 ÷ 1.1 = 140 cm
Amy's original height = 150.8 ÷ 1.04 = 145 cm
In 2013 Amy was taller than Elian by 5 cm.

Now try this

1 Shereen buys a pair of shoes in a sale where all the items are marked '40% off'.
  She pays €27 for the shoes.
  What price were the shoes originally?   **(3 marks)**

2 Jared bought a house in 2010.
  By 2012 his house had increased in value by 8%. The new value of Jared's house is €237 600. How much did Jared pay for his house?   **(3 marks)**

Worked solution video

Repeated percentage change

You need to be able to apply **repeated** percentage change. This is quickest if you use **multipliers** and **index notation**.

Compound interest

If you leave your money in a bank account it will earn compound interest.

Oliver invests HK$40 000 at a compound interest rate of 3% per annum.

'Per annum' means 'per year'

This table shows the amount of interest he earns each year.

Year	Balance (HK$)	Interest earned (HK$)
1	40 000	1200
2	41 200	1236
3	42 436	1273.08
	43 709.08	3709.08

For year 2 you have to calculate 3% of HK$41 200

After 3 years the total interest earned will be HK$3709.08 and the balance will be HK$43 709.08

Using index notation

This table uses a multiplier to work out the balance of Oliver's account at the end of each year.

Year	Balance (HK$)
1	40 000 × 1.03 = 41 200
2	41 200 × 1.03 = 42 436
3	42 436 × 1.03 = 43 709.08

You can use indices to work out the final balance after 3 years more easily.

Balance after 3 years

$$= HK\$40\,000 \times 1.03 \times 1.03 \times 1.03$$
$$= HK\$40\,000 \times 1.03^3$$
$$= HK\$43\,709.08$$

For a reminder about working with indices look at pages 2 and 3.

Worked example

At the start of an experiment, a petri dish contains 5000 cells. The number of cells in the petri dish increases by 20% each day.

Calculate the number of cells in the petri dish at the end of 4 days.   **(2 marks)**

$5000 \times 1.2^4 = 10\,368$

You can use this rule to calculate a repeated percentage change.

Final amount = (starting amount) × (multiplier)^n

n is the number of times the change is made. In this example, n is the number of days.

Repeated decrease

If a question involves **depreciation** or **decay** you might need to work out a repeated percentage decrease.

This car **depreciates** in value by 8% each year.

The multiplier for an 8% decrease is 0.92

After 3 years the car is worth:

$\$15\,000 \times 0.92 \times 0.92 \times 0.92$
$= \$15\,000 \times 0.92^3 = \$11\,680.32$

Worked solution video

Now try this

William invests £5000 in a savings account. He is paid 3% per annum compound interest.
(a) How much will William have in his savings account after 2 years?

    (2 marks)
(b) William needs £5600 to buy a car. Calculate the number of years William would need to leave his money in the account to save up this amount.   **(2 marks)**

Upper and lower bounds

Upper and lower bounds are a measure of accuracy. For example, the width of a postcard is given as 8 cm to the nearest cm.

```
       lower          upper
       bound          bound
   ├────┼────┼────┼────┤
  7 cm 7.5 cm 8 cm 8.5 cm 9 cm
        ←────────→
```

The actual width of the postcard could be anything between 7.5 cm and 8.5 cm.

7.5 cm is called the **lower bound**.

8.5 cm is called the **upper bound**.

Using upper and lower bounds in calculations

To find the overall upper and lower bounds of the answer to a calculation use these rules.

	+	−	×	÷
Overall upper bound	UB + UB	UB − LB	UB × UB	UB ÷ LB
Overall lower bound	LB + LB	LB − UB	LB × LB	LB ÷ UB

Overall lower bound of *a* + *b* = lower bound of *a* + lower bound of *b*

Worked example   Aiming higher

A roll of ribbon is 150 cm long, correct to 2 significant figures.
A 21-cm piece of ribbon is cut off the roll, correct to the nearest cm.
Calculate the lower bound, in cm, for the amount of ribbon remaining on the roll. **(3 marks)**

	Lower bound	Upper bound
Length of ribbon	145 cm	155 cm
Length of piece cut off	20.5 cm	21.5 cm

Lower bound of remaining length
= 145 − 21.5
= 123.5 cm

If you're answering questions about upper and lower bounds, it's a good idea to write out the upper bound and lower bound for **all values** given in the question before you start. To work out the **lower bound** for $a - b$ you need to use the **lower bound** for a and the **upper bound** for b.

Different values might be given to **different degrees of accuracy**. The length of the roll is correct to 2 significant figures, or to the nearest **10 cm**. The length of the piece cut off is correct to the nearest **cm**. Be really careful when you're working out your upper and lower bounds.

Now try this   Aiming higher

1   The area of a rectangle is 320 cm². The length of the rectangle is 22 cm. Both values are correct to 2 significant figures. Calculate the lower bound for the width of the rectangle. Show your working clearly. **(3 marks)**

2   Correct to 2 decimal places, the volume of a solid cube is 3.37 m³. Calculate the upper bound for the surface area of the cube.

 (4 marks)

For a cube with edges of length x, the volume is x^3 and the surface area is $6x^2$

Standard form

Numbers in standard form have two parts.

$$7.3 \times 10^{-6}$$

This part is a number greater than or equal to 1 and less than 10

This part is a power of 10

You can use standard form to write very large or very small numbers.

$$920\,000 = 9.2 \times 10^5$$

Numbers greater than 10 have a positive power of 10

$$0.007\,03 = 7.03 \times 10^{-3}$$

Numbers less than 1 have a negative power of 10

Counting decimal places

You can count decimal places to convert between numbers in standard form and ordinary numbers.

3 jumps

$$7\,9\,0\,0 = 7.9 \times 10^3$$

7900 > 10
So the power is positive

4 jumps

$$0.0\,0\,0\,3\,5 = 3.5 \times 10^{-4}$$

0.00035 < 1
So the power is negative

Be careful!
Don't just count zeros to work out the power.

Worked example

(a) Write 1 630 000 in standard form.   **(1 mark)**

1.63×10^6

(b) Write 4.2×10^{-3} as an ordinary number.   **(1 mark)**

0.0042

> Count the number of decimal places you need to move to get a number between 1 and 10. 1 630 000 is bigger than 10 so the power will be positive.

Using a calculator

You can enter numbers in standard form using the $\boxed{\times 10^x}$ key.

To enter 3.7×10^{-6} press

$\boxed{3}\,\boxed{.}\,\boxed{7}\,\boxed{\times 10^x}\,\boxed{(-)}\,\boxed{6}$

If you are using a calculator with numbers in standard form it is a good idea to put brackets around each number.

For part (b) you would enter:

$\boxed{(}\,\boxed{1}\,\boxed{.}\,\boxed{9}\,\boxed{\times 10^x}\,\boxed{4}\,\boxed{)}\,\boxed{x^2}\,\boxed{=}$

Worked example

A and B are standard form numbers.

$A = 1.9 \times 10^4 \quad B = 4.2 \times 10^5$

Calculate, giving your answers in standard form:

(a) $A + B$ **(1 mark)**

$(1.9 \times 10^4) + (4.2 \times 10^5) = 439\,000$
$= 4.39 \times 10^5$

(b) A^2 **(1 mark)**

$(1.9 \times 10^4)^2 = 3.61 \times 10^8$

Now try this

The mass of an empty Airbus A380 is 2.77×10^5 kg.

(a) Write 2.77×10^5 as an ordinary number.   **(1 mark)**

On take-off, an A380 carries 2.4×10^5 kg of fuel plus passengers and crew with a total mass of 3.41×10^4 kg.

(b) Calculate the total take-off weight of the A380. Give your answer in standard form.   **(2 marks)**

Problem-solving practice 1

In your International GCSE Maths exams you will need to demonstrate **problem-solving** and **reasoning skills**. If you come across a tricky or unfamiliar question in your exam, you can try some of these strategies:

- ☑ Sketch a diagram to see what is going on.
- ☑ Try the problem with smaller or easier numbers.
- ☑ Plan your strategy before you start.
- ☑ Write down any formulae you might be able to use, or check the formulae sheet for a clue.
- ☑ Use x or n to represent an unknown value.

---

1 The diagram shows two types of plastic building block.

 A 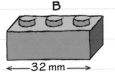 B

Worked solution video

←24 mm→   ←32 mm→

Block **A** is 24 mm long.

Block **B** is 32 mm long.

Jeremy joins some type **A** blocks together to make a straight row.

He then joins some type **B** blocks together to make a straight row of the same length.

Write down the shortest possible length of this row. **(4 marks)**

Factors and primes page 1

You have to use a whole number of building blocks in each row, so the length of each row has to be a multiple of the length of one block. The answer will be the lowest common multiple of 24 and 32. You can't get all the marks just by writing down the answer. You need to show how you found the answer clearly and neatly.

TOP TIP

If you're not sure how to start, draw a sketch. This might help you see that the lengths are multiples of 24 and 32.

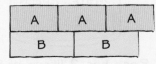

A	A	A
B		B

---

2 Susan has 2 cats.

Each cat is fed $\frac{3}{8}$ kg of cat food each day.

Susan buys cat food in bags.

Each bag weighs 14 kg.

For how many days can Susan feed the 2 cats from 1 bag of cat food?

You must show **all** your working. **(5 marks)**

Fractions page 5

There are lots of steps in this question so make sure you keep track of your working and write it down clearly.

TOP TIP

Write words with each calculation to explain what you are doing.

---

3 Prove that the recurring decimal 0.9̇2̇8̇ can be written as $\frac{919}{990}$

(3 marks)

 Worked solution video

Recurring decimals page 7

 Aiming higher

Start by writing $x = 0.9282828...$ If you work out $1000x$ and $10x$ you can keep all your working in whole numbers.

TOP TIP

If a question says 'Show that ...' or 'Prove that ...' you have to show every step of your working clearly and neatly.

Problem-solving practice 2

 This item appeared in a newspaper.

> **Cow produces 3% more milk**
>
> A farmer found that when his cow listened to classical music, the milk it produced increased by 3%.
> This increase of 3% represented 0.72 litres of milk.

Calculate the amount of milk produced by the cow when it listened to classical music.

(3 marks)

Proportion page 11
Reverse percentages page 13

When the cow listened to classical music, it produced 103% of the milk it produced originally. You know that 3% represents 0.72 litres. Use this information to work out what 103% represents.

TOP TIP

You can sometimes solve percentage problems by working out what 1% represents.

 A paint manufacturer mixes white and green paint in two different ratios to make different shades.

APPLE WHITE
RATIO
WHITE : GREEN
5 : 2

SUMMER FIELDS
RATIO
WHITE : GREEN
1 : 4

Rosa has 2 litres of Summer Fields. How much extra white paint should she add to turn it into Apple White? **(4 marks)**

Ratio page 10

Start by working out how much white and green paint are contained in the 2 litres of Summer Fields that Rosa already has. Her new mixture will contain the same amount of **green** paint. You can use the ratio to work out what the **total** amount of white paint should be in the new mixture.

TOP TIP

Plan your strategy before you start – you will save time and your working will be much clearer.

 The diagram shows a wooden planting box in the shape of a cuboid.

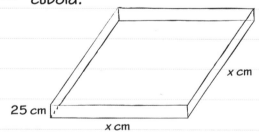

25 cm
x cm
x cm

The volume of the box is 810 000 cm³ correct to 2 significant figures.

The depth of the box is 25 cm, to the nearest cm.

The box has a square base with sides of length x cm.

Find the lower bound for x. Give your answer correct to 3 significant figures.

(4 marks)

Upper and lower bounds page 15
Volumes of 3D shapes page 79

Aiming higher

Complete this table showing the upper and lower bounds for each measurement before you start:

	25 cm	810 000 cm
Upper	25.5 cm	
Lower		805 000 cm

You are **dividing** the volume by the depth to work out x^2. Choose the values you use carefully to make the answer as **small** as possible.

TOP TIP

When answering questions about upper and lower bounds, it's a good idea to write out the upper and lower bounds for all the values before you start.

Algebraic expressions

You need to be able to work with algebraic expressions confidently. For a reminder about using the index laws with **numbers** have a look at pages 2 and 3.

 1 You can use the **index laws** to simplify algebraic expressions.

$a^m \times a^n = a^{m+n}$

$x^4 \times x^3 = x^{4+3} = x^7$

$\dfrac{a^m}{a^n} = a^{m-n}$

$m^8 \div m^2 = m^{8-2} = m^6$

$(a^m)^n = a^{mn}$

$(n^2)^4 = n^{2 \times 4} = n^8$

2 You can square or cube a whole expression.

$(4x^3y)^2 = (4)^2 \times (x^3)^2 \times (y)^2$
$\qquad\qquad = 16x^6y^2$

$16 = (4)^2$

$(x^3)^2 = x^{3 \times 2} = x^6$

You need to square everything inside the brackets.

Remember that if a letter appears on its own then it has the power 1.

3 Algebraic expressions may also contain negative and fractional indices.

$a^{-m} = \dfrac{1}{a^m}$

$(c^2)^{-3} = c^{2 \times -3} = c^{-6} = \dfrac{1}{c^6}$

$a^{\frac{1}{n}} = \sqrt[n]{a}$

$(8p^3)^{\frac{1}{3}} = (8)^{\frac{1}{3}} \times (p^3)^{\frac{1}{3}}$
$\qquad\qquad = \sqrt[3]{8} \times p^{3 \times \frac{1}{3}}$
$\qquad\qquad = 2p$

One at a time

When you are **multiplying** expressions:

1. Multiply any number parts first.

2. Add the powers of each letter to work out the new power.

$$6p^2q \times 3p^3q^2 = 18p^5q^3$$

$6 \times 3 = 18$

$p^2 \times p^3 = p^{2+3} = p^5$

$q \times q^2 = q^{1+2} = q^3$

When you are **dividing** expressions:

1. Divide any number parts first.

2. Subtract the powers of each letter to work out the new power.

$12 \div 3 = 4$

$b^3 \div b^2 = b^{3-2} = b$

$$\dfrac{12a^5b^3}{3a^2b^2} = 4a^3b$$

$a^5 \div a^2 = a^{5-2} = a^3$

Worked example

Simplify fully

(a) $m^2 + m^2 + m^2 + m^2$ **(1 mark)**

$4m^2$

(b) $(x^3)^3$ **(1 mark)**

x^9

(c) $\dfrac{4y^2 \times 3y^7}{6y}$ **(2 marks)**

$\dfrac{4y^2 \times 3y^7}{6y} = \dfrac{12y^9}{6y} = 2y^8$

(a) This is four lots of m^2, so you write it as $4 \times m^2$ or $4m^2$

(b) Use $(a^m)^n = a^{mn}$

(c) Start by simplifying the top part of the fraction. Do the number part first then the powers. Use $a^m \times a^n = a^{m+n}$

Next divide the expressions. Divide the number part, then divide the indices using $\dfrac{a^m}{a^n} = a^{m-n}$

Worked solution video

Now try this

1 Simplify $(h^2)^6$ **(1 mark)**

2 Simplify fully

(a) $(2a^5b)^4$ **(2 marks)**

(b) $5x^4y^2 \times 3x^3y^7$ **(2 marks)**

(c) $18d^8g^{10} \div 6d^2g^5$ **(2 marks)**

 Aiming higher

3 (a) Simplify $(16p^{10})^{\frac{1}{2}}$ **(2 marks)**

(b) Simplify $(64x^9y^2)^{-\frac{1}{3}}$ **(2 marks)**

Apply the power outside the brackets to everything inside the brackets.

Expanding brackets

Expanding or multiplying out brackets is a key algebra skill.

You have to multiply the expression outside the bracket by everything inside the bracket.

$4n \times n^2 = 4n^3$

$$4n(n^2 + 2) = 4n^3 + 8n$$

$4n \times 2 = 8n$

'Expand and simplify' means 'multiply out and then collect like terms'.

Golden rule

When you expand, you need to be careful with negative signs in front of the bracket.

Negative signs belong to the term to their right.

$-2 \times x$ $\quad -2 \times -y$

$$x - 2(x - y) = x - 2x + 2y$$
$$= -x + 2y$$

Multiply out the brackets first and then collect like terms if possible.

You can use the **grid method** to expand two brackets.

$(x + 7)(x - 5) = x^2 - 5x + 7x - 35$
$$= x^2 + 2x - 35$$

Remember to collect like terms if possible.

	x	−5
x	x^2	$-5x$
7	$7x$	-35

The negative sign belongs to the 5.

You need to write it in your grid.

Or

You can use the acronym FOIL to expand two brackets.

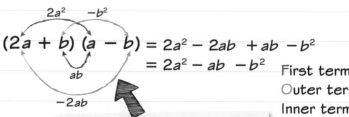

$2a^2$ $\quad -b^2$

$(2a + b)(a - b) = 2a^2 - 2ab + ab - b^2$
$$= 2a^2 - ab - b^2$$

ab

$-2ab$

Some people remember this as a 'smiley face'.

First terms
Outer terms
Inner terms
Last terms

(a) Expand and simplify $(3p - 4)^2$ **(2 marks)**

$(3p - 4)^2 = (3p - 4)(3p - 4)$
$$= 9p^2 - 12p - 12p + 16$$
$$= 9p^2 - 24p + 16$$

	3p	−4
3p	$9p^2$	$-12p$
−4	$-12p$	16

(b) Expand and simplify $x(x - 2)(x + 3)$ **(2 marks)**

$x(x - 2)(x + 3) = x(x^2 + 3x - 2x - 6)$
$$= x(x^2 + x - 6)$$
$$= x^3 + x^2 - 6x$$

Be careful with the negative signs:
$-4 \times -4 = 16$ $3p \times -4 = -12p$

You might have to multiply three factors together. Start by expanding $(x - 2)(x + 3)$ and remember to write brackets around the whole expansion. Then multiply the term outside the brackets, x, by **every term** inside the brackets.

Check it!

Try an easy value, like $x = 5$
$x(x - 2)(x + 3) = 5 \times 3 \times 8 = 120$
$x^3 + x^2 - 6x = 125 + 25 - 30 = 120$ ✓

Worked solution video

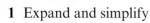

1 Expand and simplify

(a) $(x + 5)(x - 1)$ **(2 marks)**

(b) $(p - 6)^2$ **(2 marks)**

2 Expand and simplify

(a) $x(x + 3)(x + 9)$ **(2 marks)**

(b) $(n + 5)(n + 3)^2$ **(3 marks)**

$(n + 5)(n + 3)^2 = n(n + 3)^2 + 5(n + 3)^2$

Factorising

Factorising is the opposite of expanding brackets:

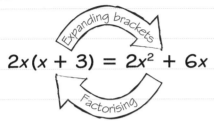

$$2x(x + 3) = 2x^2 + 6x$$

You need to look for the **largest factor** you can take out of every term in the expression.

$$10a^2 + 5ab = 5(2a^2 + ab)$$

This expression has only been **partly factorised**.

$$10a^2 + 5ab = 5a(2a + b)$$

This expression has been **completely factorised**.

Factorising $x^2 + bx + c$

You need to write the expression with **two brackets**.

You need to find two numbers which add up to 7... $5 + 2 = 7$

$$x^2 + 7x + 10 = (x + 5)(x + 2)$$

... and multiply to make 10 $5 \times 2 = 10$

When factorising $x^2 + bx + c$, use this table to help you find the two numbers:

b	c	Factors
Positive	Positive	Both numbers positive
Positive	Negative	Bigger number positive and smaller number negative
Negative	Negative	Bigger number negative and smaller number positive
Negative	Positive	Both numbers negative

Factorising $ax^2 + bx + c$

If the coefficient of x^2 is more than 1, find two numbers that add up to b and multiply to make ac. For example:

$$8x^2 + 22x + 15$$

$ac = 120 = 12 \times 10$, and $12 + 10 = 22$

So $8x^2 + 22x + 15$

$$= 8x^2 + 10x + 12x + 15$$

$$= 2x(4x + 5) + 3(4x + 5)$$

$$= (4x + 5)(2x + 3)$$

Difference of two squares

You can factorise expressions that are written as

$$(\text{something})^2 - (\text{something else})^2$$

Use this rule:

$$a^2 - b^2 = (a + b)(a - b)$$

$$x^2 - 36 = x^2 - (6)^2$$

$$= (x + 6)(x - 6)$$

36 is a square number.

$36 = 6^2$ so $a = x$ and $b = 6$

Worked example

Factorise $x^2 - x - 20$ **(2 marks)**

Factor pairs of 20:
1 and 20, 2 and 10, 4 and 5
$x^2 - x - 20 = (x + 4)(x - 5)$
Check:
$(x + 4)(x - 5) = x^2 - 5x + 4x - 20$
$= x^2 - x - 20$ ✔

The answer will have **two sets of brackets**.

The last term is negative, so the brackets will have one + sign and one − sign. With any factorisation, the safest thing to do is to **check your answer** by expanding the brackets.

Now try this

1. Factorise
   (a) $x^2 + 12x + 20$ **(2 marks)**
   (b) $x^2 - 3x - 10$ **(2 marks)**

2. Factorise fully
   (a) $12g + 3g^2$ **(2 marks)**
   (b) $p^2 - 15p + 14$ **(2 marks)**
   (c) $6x^2 - 8xy$ **(2 marks)**

3. Factorise fully
   (a) $4ma - 24m^2a$ **(2 marks)**
   (b) $p^2 - 64$ **(1 mark)**

4. Factorise $3x^2 - 8x + 4$ **(2 marks)**

Worked solution video

Linear equations 1

To solve a linear equation you need to get the letter on its own on one side.
It is really important to write your working **neatly** when you are solving equations.

$$5x + 3 = 18 \quad (-3)$$

Every line of working should have an equals sign in it.

$$5x = 15 \quad (÷ 5)$$

$$x = 3$$

Write down the operation you are carrying out. Remember to do the same thing to both sides of the equation.

Start a new line for each step.
Do one operation at a time.

Line up the equals signs.

Letter on both sides?

To solve an equation you have to get the letter on its own on one side of the equation.

Start by collecting like terms so that all the letters are together.

$$2 - 2x = 26 + 4x \quad (+ 2x)$$

$$2 = 26 + 6x \quad (- 26)$$

$$-24 = 6x \quad (÷ 6)$$

$$-4 = x$$

You can write your answer as

$-4 = x$ or as $x = -4$

Equations with brackets

Always start by multiplying out the brackets then collecting like terms.

For a reminder about multiplying out brackets have a look at page 20.

$$19 = 8 - 2(5 - 3y)$$

$$19 = 8 - 10 + 6y$$

$$19 = -2 + 6y \quad (+ 2)$$

$$21 = 6y \quad (÷ 6)$$

$$\frac{21}{6} = y$$

$$y = \frac{7}{2} \text{ or } 3\frac{1}{2} \text{ or } 3.5$$

Your answer can be written as a fraction or decimal.

Worked example

Solve $7r + 2 = 5(r - 4)$ **(3 marks)**

$$7r + 2 = 5r - 20 \quad (- 5r)$$

$$2r + 2 = -20 \quad (- 2)$$

$$2r = -22 \quad (÷ 2)$$

$$r = -11$$

Multiply out the brackets then collect all the terms in r on one side. You need to write down each step of your working clearly.

Check it!

Substitute $r = -11$ into each side of the equation.

Left-hand side: $7(-11) + 2 = -75$

Right-hand side: $5(-11 - 4) = -75$ ✓

Now try this

1 Solve
   (a) $5w - 17 = 2w + 4$ **(3 marks)**
   (b) $2(x + 11) = 20$ **(3 marks)**

2 Solve
   (a) $6y - 9 = 2(y - 8)$ **(3 marks)**
   (b) $4m - 2(m - 3) = 7m - 14$ **(3 marks)**

You could start by expanding the brackets, or by dividing both sides by 2.

Expand the brackets then collect all the m terms on one side of the equation.

Linear equations 2

Equations with fractions

When you have an equation with fractions, you need to get rid of any fractions before solving. You can do this by multiplying every term by the lowest common multiple (LCM) of the denominators.

$$\frac{x}{3} + \frac{x-1}{5} = 11 \quad (\times 15)$$

The LCM of 3 and 5 is 15.

$$\frac{^5\cancel{15}x}{\cancel{3}_1} + \frac{^3\cancel{15}(x-1)}{\cancel{5}_1} = 165$$

Cancel the fractions. There is more about simplifying algebraic fractions on page 52.

$$5x + 3x - 3 = 165$$
$$8x - 3 = 165 \quad (+3)$$
$$8x = 168 \quad (\div 8)$$
$$x = 21$$

Multiplying by an expression

You might have to multiply by an expression to get rid of the fractions.

$$\frac{20}{n-3} = -5 \quad (\times(n-3))$$
$$20 = -5(n-3)$$

Worked example

Solve $\dfrac{29-x}{4} = x + 5$ **(3 marks)**

$$\frac{4(29-x)}{4} = 4(x+5)$$
$$29 - x = 4(x+5)$$
$$29 - x = 4x + 20 \quad (+x)$$
$$29 = 5x + 20 \quad (-20)$$
$$9 = 5x \quad (\div 5)$$
$$\frac{9}{5} = x$$

Eliminate fractions **before** you start solving the equation. You can do this by multiplying both sides of the equation by 4.
Use brackets to show that you are multiplying everything by 4.
$4(x+5)$ ✓ $4x + 5$ ✗
Multiply out the brackets, then solve the equation normally. Remember that your answer could be a fraction.

Top tip!

It's OK to leave the answer to an equation as an improper fraction. Don't waste time converting to mixed numbers or decimals.

Writing your own equations

You can find unknown values by writing and solving equations.

$$4(x-1) = 3x + 3 \qquad \frac{5}{n} + \frac{5}{n} + 2 + 2 = 20$$

$4(x-1)$ cm $(3x+3)$ cm

$\frac{5}{n}$ m   Perimeter = 20 m   2 m

Now try this

1 Solve

  (a) $\dfrac{25 - 3w}{4} = 10$ **(3 marks)**

  (b) $5x - 10 = \dfrac{18 - x}{3}$ **(3 marks)**

2 Solve

  (a) $\dfrac{2y}{3} + \dfrac{y - 4}{2} = 5$ **(3 marks)**

  (b) $\dfrac{3m - 1}{4} - \dfrac{2m + 4}{3} = 1.5$ **(3 marks)**

Formulae

A **formula** is a mathematical rule.

You can write formulae using algebra.

This label shows a formula for working out the cooking time of a chicken.

FREE-RANGE CHICKEN		
WEIGHT (KG)	PRICE PER KG	COOKING INSTRUCTIONS
1.8	€3.95	Cook at 170 °C for 45 minutes per kg plus half an hour

You can write this formula using algebra as

$T = 45w + 30$, where T is the cooking time in minutes and w is the weight in kg.

In the description of each variable, you must give the units.

If T was the cooking time in hours then this formula would give you a very crispy chicken!

Worked example

This formula is used in physics to calculate distance:

$D = ut - 5t^2$

$u = 14$ and $t = -3$

Work out the value of D.   **(2 marks)**

$D = (14)(-3) - 5(-3)^2$

$\quad = (14)(-3) - 5(9)$

$\quad = -42 - 45$

$\quad = -87$

Substitute the values for u and t into the formula.

If you use brackets then you're less likely to make a mistake. This is really important when there are negative numbers involved.

Remember **BIDMAS** for the correct order of operations. You need to do:

Indices → Multiplication → Subtraction

Don't try to do more than one operation on each line of working.

Worked example

The area of this shape is A cm².

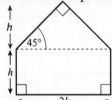

45°

2h

Everything in red is part of the answer.

Write a formula for A in terms of h.
Give your answer in its simplest form. **(3 marks)**

$A = 2h \times h + \frac{1}{2} \times 2h \times h$

$\quad = 2h^2 + h^2$

$A = 3h^2$

Problem solved!

You are only given the dimensions of the rectangle. You need to **infer** (i.e. guess based on information) the height of the triangle. The angle of the slope is 45° and the base of the triangle is $2h$ so the height of the triangle must be h. Write this dimension on your diagram. Use these dimensions to write an expression for the area of the triangle. Remember to write '$A =$'. If you only write '$3h^2$' it is an **expression**, not a **formula**.

You will need to use problem-solving skills throughout your exam – **be prepared!**

Now try this

The perimeter of this shape can be calculated using the formula

$$P = \frac{4(a^2 + ab + b^2)}{a + b}$$

Find the value of P when $a = 4.3$ cm and $b = 2.9$ cm.
Give your answer correct to 3 significant figures.

   (2 marks)

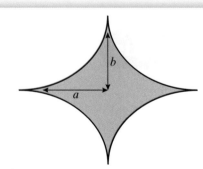

Arithmetic sequences

An **arithmetic or linear sequence** is a sequence of numbers where the difference between consecutive terms is **constant**. In your exam, you might need to work out the nth term of a sequence. Look at this example, which shows you how to do it in four steps.

Worked example

Here is a sequence.

$1 \boxed{+4} 5 \boxed{+4} 9 \boxed{+4} 13 \boxed{+4} 17$

Work out a formula for the nth term of the sequence. **(2 marks)**

Write in the difference between each term.

Here is a sequence.

Zero term
$-3 \quad 1 \boxed{+4} 5 \boxed{+4} 9 \boxed{+4} 13 \boxed{+4} 17$

Work out a formula for the nth term of the sequence. **(2 marks)**

Find the imaginary term before the first term by working backwards. You can call this the **zero term**.

Here is a sequence.

Zero term
$-3 \quad 1 \boxed{+4} 5 \boxed{+4} 9 \boxed{+4} 13 \boxed{+4} 17$

Work out a formula for the nth term of the sequence. **(2 marks)**

nth term = difference $\times$ n + zero term

Write down the formula for the nth term. **Remember** this formula for the exam.

Here is a sequence.

Zero term
$-3 \quad 1 \boxed{+4} 5 \boxed{+4} 9 \boxed{+4} 13 \boxed{+4} 17$

Work out a formula for the nth term of the sequence. **(2 marks)**

nth term = difference $\times$ n + zero term
nth term = $4n - 3$

You can also use a formula to work out the nth term of an arithmetic sequence. There is more about this on page 51.

Is 99 in this sequence?

You can use the nth term to check whether a number is a term in the sequence.

The value of n in your nth term has to be a **positive** whole number.

Try some different values of n:

$n = 25 \rightarrow 4n - 3 = 97$

$n = 26 \rightarrow 4n - 3 = 101$

You can't use a value of n between 25 and 26 so 99 is **not** a term in the sequence.

Generating sequences

You can work out the terms of a sequence by substituting the term number into the nth term. Here are two examples:

nth term	$n^2 + 10$	$\sqrt{3}^{\,n}$
1st term	$1^2 + 10 = 11$	$\sqrt{3}^1 = \sqrt{3}$
2nd term	$2^2 + 10 = 14$	$\sqrt{3}^2 = 3$
3rd term	$3^2 + 10 = 19$	$\sqrt{3}^3 = 3\sqrt{3}$
⋮	⋮	⋮
8th term	$8^2 + 10 = 74$	$\sqrt{3}^8 = 81$

Now try this

Here are the first five terms of an arithmetic sequence.
3, 7, 11, 15, 19

(a) Write down an expression for the nth term. **(2 marks)**

Karen says that 89 is a term in the sequence.

(b) Is she right? Give reasons for your answer. **(2 marks)**

Method 1
Write down some terms in the sequence that are close to 89.

Method 2
Set the nth term equal to 89 and solve the equation.

Worked solution video

Straight-line graphs 1

Here are two things you need to know about straight-line graphs.

 If an equation is in the form $y = mx + c$, its graph will be a straight line.

$$y = -\tfrac{1}{2}x + 5$$

This number tells you the gradient of the graph.

The y-intercept of the graph is at $(0, 5)$.

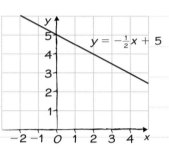

$y = -\tfrac{1}{2}x + 5$

The gradient of the graph is $-\tfrac{1}{2}$.

This means that for every unit you go across, you go half a unit down.

 Use a table of values to draw a graph.

$y = 2x + 1$

x	-1	0	1	2
y	-1	1	3	5

$\rightarrow y = 2 \times 2 + 1 = 5$

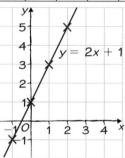

$y = 2x + 1$

Choose simple values of x and substitute them into the equation to find the values of y.

Plot the points on your graph and join them with a straight line.

Worked example

On the grid, draw the graph of $x + y = 4$ for values of x from -2 to 5

(3 marks)

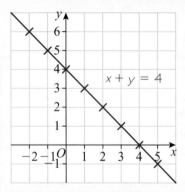

$x + y = 4$

Everything in red is part of the answer.

x	-2	-1	0	1	2	3	4	5
y	6	5	4	3	2	1	0	-1

Finding equations

If you have a graph you can find its equation by working out the gradient and looking at the y-intercept.

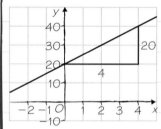

Draw a triangle to find the gradient.

Gradient $= \dfrac{20}{4} = 5$

The y-intercept is $(0, 20)$.

Put your values for gradient, m, and y-intercept, c, into the equation of a straight line, $y = mx + c$.

The equation is $y = 5x + 20$

You can rearrange the equation of this graph into the form $y = mx + c$ so it is a **straight line**.

$x + y = 4$ $(-x)$

     $y = -x + 4$ $m = -1$ and $c = 4$

The gradient is -1 and the y-intercept is at $(0, 4)$. You could use this information to draw the graph, but it's safer to make a table of values. Make sure you plot **at least three** points, then join them with a straight line **using a ruler**. Make sure you read carefully from the scale on each axis – don't just count squares!

Now try this

Find the equation of the straight line.

(3 marks)

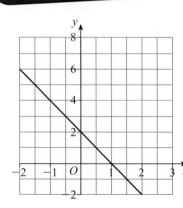

Use $y = mx + c$. Draw a triangle to find the gradient, m. The graph slopes down so m will be **negative**.

Straight-line graphs 2

On page 26 you revised how to find an equation by looking at a graph. You can also use **algebra** to find the equation. This is usually a more reliable and accurate method.

1 **Given one point and the gradient**

	Gradient 2, passing through point (3, 7)
Substitute the gradient for m in $y = mx + c$	

$y = 2x + c$
$7 = 2 \times 3 + c$

Substitute the x- and y-values given into the equation

$7 = 6 + c \quad (-6)$
$c = 1$

Solve the equation to find c

$y = 2x + 1$

Write out the equation

2 **Given two points**

Draw a sketch showing the two points

Passing through points (8, 20) and (10, 30)

(10, 30)
10
(8, 20)
2

Work out the gradient of the line using a triangle

Gradient $= \dfrac{10}{2} = 5$

Use method 1 (on the left) and one of the points given to find the equation

$y = 5x + c$
$30 = 5 \times 10 + c$
$c = -20$
$y = 5x - 20$

Worked example

A line passes through the points with coordinates (1, 5) and (2, 7).
Find an equation of the line.

(3 marks)

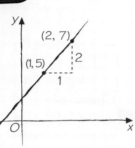

Gradient, $m = \dfrac{2}{1} = 2$

Equation of line: $y = mx + c \rightarrow y = 2x + c$

For point (1, 5), $x = 1$, $y = 5$

Substitute these values into the equation:
$5 = 2 + c \rightarrow c = 3$

The equation is $y = 2x + 3$

Positive or negative?

If the line slopes **down** then the gradient is **negative**.

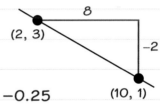
8
(2, 3)
-2
(10, 1)

Gradient $= \dfrac{-2}{8} = -0.25$

Follow these steps:
1. Draw a **sketch** of the straight line.
2. Draw a **triangle** to find the gradient. This is your value for m. Put it into the equation for the line.
3. Use the x- and y-values of one of the points on the line to write an equation.
4. **Solve** your equation to find c.

Now try this

Worked solution video

1 A straight line has gradient 6 and passes through the point (5, 10).
   Find an equation of the line. **(2 marks)**

2 Find the equation of a straight line that passes through the points (3, 5) and (7, 13). **(3 marks)**

3 A straight line passes through the points with coordinates $(-3, -2)$, $(1, 6)$ and $(k, 16)$.
   Work out the value of k. You must show all your working. **(4 marks)**

Find the gradient, m, then substitute $x = 3$ and $y = 5$ into $y = mx + c$. Solve an equation to find the value of c.

Parallel and perpendicular

Parallel lines have the same gradient.

These three lines all have a gradient of 1

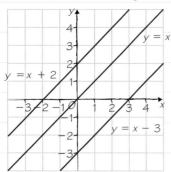

Perpendicular means at right angles.

If a line has gradient m then any line perpendicular to it will have gradient $-\dfrac{1}{m}$

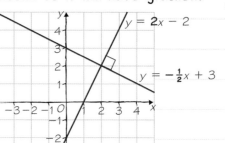

Worked example

Aiming higher

A line L passes through the points $(-3, 6)$ and $(5, 4)$.

Another line, P, is perpendicular to L and passes through the point $(0, -7)$. Find the equation of line P. **(3 marks)**

Gradient of line L

$= \dfrac{-2}{8} = \dfrac{-1}{4}$

Gradient of line P

$= \dfrac{-1}{\frac{-1}{4}} = 4$

P passes through $(0, -7)$

Equation of P is: $y = 4x - 7$

1. Draw a sketch to find the gradient of line L.
2. The line slopes down so the gradient is negative.
3. Use $-\dfrac{1}{m}$ to calculate the gradient of P. If m is a fraction, you can just find its reciprocal and change the sign.
4. You know P passes through $(0, -7)$. Use $m = 4$ and $c = -7$ to write the equation of line P.

Check it!

If two lines are perpendicular, the product of their gradients is -1: $-\dfrac{1}{4} \times 4 = -1$ ✔

Midpoints

A **line segment** is a short section of a straight line.

You can find the **midpoint** of a line segment if you know the coordinates of the ends.

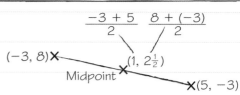

Coordinates of midpoint = (average of x-coordinates, average of y-coordinates)

Now try this

1 Line A has equation $y = \frac{1}{2}x + 1$

Line B is parallel to line A.

Work out an equation for line B. **(3 marks)**

Aiming higher

2 Point C has coordinates $(3, 4)$ and point D has coordinates $(9, 12)$.

(a) Find the equation of the line CD. **(2 marks)**

(b) Find the equation of the perpendicular bisector of the line CD. **(2 marks)**

Quadratic graphs

Quadratic equations contain an x^2 term. Quadratic equations have **curved** graphs. You can draw the graph of a quadratic equation by completing a table of values.

The **turning point** is the point where the direction of the curve changes.

Worked example

(a) Complete the table of values for $y = 4x - x^2$. **(2 marks)**

x	-1	0	1	2	3	4	5
y	-5	0	3	4	3	0	-5

(b) On the grid, draw the graph of $y = 4x - x^2$. **(2 marks)**
(c) Write down the coordinates of the turning point.

           (1 mark)

(2, 4)

When $x = -1$: $4 \times (-1) - (-1)^2 = -4 - 1 = -5$
When $x = 4$: $4 \times 4 - 4^2 = 16 - 16 = 0$
Plot your points carefully on the graph and join them with a **smooth** curve.

Check it!
All the points on your graph should lie on the curve. If one of the points doesn't fit then double check your calculation.

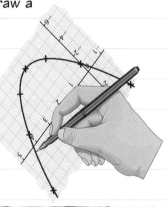

Everything in red is part of the answer.

Drawing a smooth curve

It's easier to draw a smooth curve if you turn your graph paper so your hand is **inside** the curve.

Drawing quadratic curves

- ✓ Use a sharp pencil.
- ✓ Plot the points carefully.
- ✓ Draw a smooth curve that passes through every point.
- ✓ Label your graph.
- ✓ Shape of graph will be either ∪ or ∩

Now try this

Worked solution video

(a) Complete this table of values for $y = x^2 + 2$

x	-3	-2	-1	0	1	2	3
y		6		2	3	6	

         (2 marks)

(b) Draw the graph of $y = x^2 + 2$ for $x = -3$ to $x = 3$ **(2 marks)**

(c) Use your graph to find the value of y when $x = 2.5$ **(1 mark)**

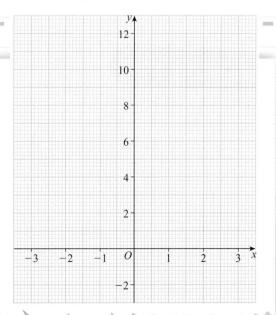

Cubic and reciprocal graphs

You might need to **draw** or **interpret** cubic and reciprocal graphs in your exam. You can use a **table of values** to draw any graph, but it helps if you know what the **general shape** of the graph is going to be.

① Cubic graphs

Graphs that contain an x^3 term and no higher powers of x are called **cubic graphs**. Here are two examples.

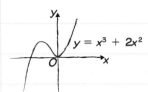
$y = x^3 + 2x^2$

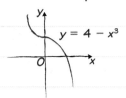
$y = 4 - x^3$

② Reciprocal graphs

Graphs of the form $y = \frac{k}{x}$ where k is a number are called **reciprocal graphs**.

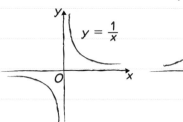
$y = \frac{1}{x}$

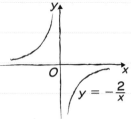
$y = -\frac{2}{x}$

The graphs get closer and closer to the x- and y-axes but never touch them.

Worked example

(a) Complete the table of values for $y = x^3 - 4x - 3$

x	-2	-1	0	1	2	3
y	-3	0	-3	-6	-3	12

(2 marks)

(b) On the grid, draw the graph of $y = x^3 - 4x - 3$ for $-2 \leqslant x \leqslant 3$ **(2 marks)**

(c) (i) Estimate the value of x when $y = 6$ **(1 mark)**

2.7

(ii) Comment on the accuracy of your estimate. **(1 mark)**

The estimate is not very accurate because it is based on reading off a graph.

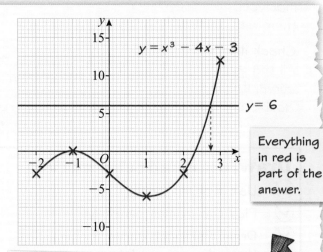

Everything in red is part of the answer.

This is a **cubic graph** with a **positive** coefficient of x. If you recognise the shape of the graph then it's easier to tell if you've plotted your coordinates correctly.

Now try this

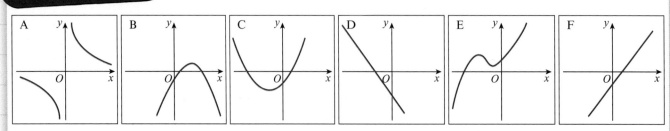

Write down the letter of the graph which could have the equation

(a) $y = -4x - 2$

(b) $y = x^3 - 3x + 4$

(c) $y = \frac{1}{x}$

(d) $y = 8x - 15 - x^2$

(e) $y = 2x - 3$

(f) $y = x^2 + 2x - 8$ **(6 marks)**

Rates of change

Distance–time graphs

A **distance–time** graph shows how distance changes with time. This distance–time graph shows Jodi's run. The shape of the graph gives you information about the journey.

A horizontal line means no movement. Jodi rested here for 15 minutes.

The gradient of the graph gives Jodi's speed.

$$\text{Gradient} = \frac{\text{distance in miles}}{\text{time in hours}} = 1.9 \div \tfrac{1}{2} = 3.8$$

Jodi was travelling at 3.8 mph on this section of the run.

This is when Jodi turned around and started heading back home.

The horizontal scale might be marked in minutes or hours. Remember that there are 60 minutes in 1 hour.

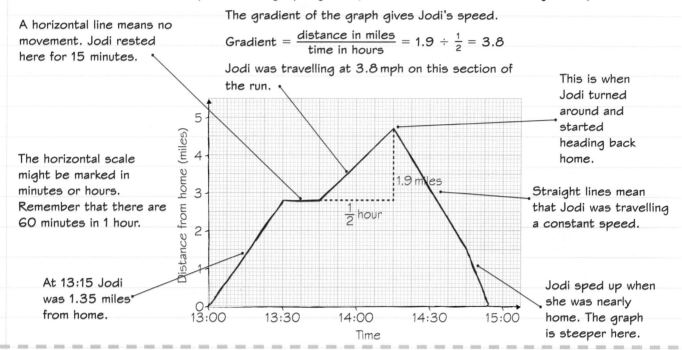

1.9 miles

½ hour

Straight lines mean that Jodi was travelling a constant speed.

At 13:15 Jodi was 1.35 miles from home.

Jodi sped up when she was nearly home. The graph is steeper here.

Rates of change

The **gradient** on a distance–time graphs tells you the **rate of change** of distance with time. This is also called **speed**. You can use graphs to find other rates of change. These garden ponds are filled with water at a constant rate. The graphs show how the depth of water in each pond changes with time.

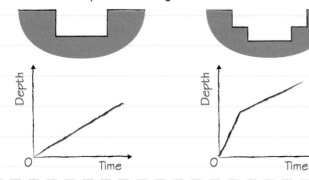

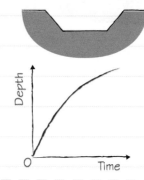

The gradient of the graph tells you the **rate of change** of the depth of water at that point. Where the pond is narrower, the rate of change will be higher.

Now try this

Li Jing rode her bike to a friend's house. She rested once on the way. She had coffee at her friend's house, then rode home.

Worked solution video

(a) How long did Li Jing spend at her friend's house?   **(1 mark)**

(b) Work out Li Jing's average speed for her return journey.   **(2 marks)**

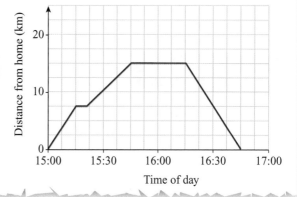

Quadratic equations

Quadratic equations can be written in the form $ax^2 + bx + c = 0$ where a, b and c are numbers.

You need to be able to **solve** a quadratic equation using factorisation.

1. **Rearrange** it into the form $ax^2 + bx + c = 0$

2. **Factorise** the left-hand side.

3. Set each factor **equal to zero** and solve to find two values of x.

For a reminder about factorising quadratic expressions have a look at page 21.

Two to watch

 When $c = 0$:
$$x^2 - 10x = 0$$
$$x(x - 10) = 0$$
Solutions are $x = 0$ and $x = 10$

 When $b = 0$ (difference of two squares):
$$9x^2 - 4 = 0$$
$$(3x + 2)(3x - 2) = 0$$
Solutions are $x = \frac{2}{3}$ and $x = -\frac{2}{3}$

Worked example

Solve $x^2 + 8x - 9 = 0$ **(3 marks)**

$$(x + 9)(x - 1) = 0$$

$$x + 9 = 0 \qquad x - 1 = 0$$
$$x = -9 \qquad x = 1$$

Follow the three steps given above.

1. The equation is already in the right form.

2. To factorise look for two numbers which add up to 8 and multiply to make -9. The numbers are 9 and -1.

3. Set each factor equal to 0 and solve.

Check it!
$(1)^2 + 8(1) - 9 = 1 + 8 - 9 = 0$ ✓
$(-9)^2 + 8(-9) - 9 = 81 - 72 - 9 = 0$ ✓

Worked example

Aiming higher

Solve $2(x + 1)^2 = 3x + 5$ **(4 marks)**

$$2(x^2 + 2x + 1) = 3x + 5$$
$$2x^2 + 4x + 2 = 3x + 5$$
$$2x^2 + x - 3 = 0$$
$$(2x + 3)(x - 1) = 0$$

$$2x + 3 = 0 \qquad x - 1 = 0$$
$$x = -\frac{3}{2} \qquad x = 1$$

When you are solving a quadratic equation you must rearrange it into the form $ax^2 + bx + c = 0$ **before** you factorise.

Be really careful if the coefficient of x^2 is bigger than 1. The two factors will look like this:

$$2x^2 + x - 3 = 0$$
$$(2x \pm \ldots)(x \pm \ldots) = 0$$

The number part of the expression is -3, so the numbers in the factors must be either -1 and 3 or 1 and -3.

Now try this

When you are solving an equation you must always show **clear algebraic working**. Don't just write down the answers without any working.

Aiming higher

Solve
(a) $m^2 - 8m + 12 = 0$ **(3 marks)**

(b) $w^2 - 36 = 5w$ **(3 marks)**

(c) $5y^2 + 37y - 24 = 0$ **(3 marks)**

(d) $8x^2 - 4 = (x - 1)^2$ **(4 marks)**

The quadratic formula

This is how the quadratic formula will appear on the formula sheet in your exam.

The Quadratic Equation

The solutions of $ax^2 + bx + c = 0$ where $a \neq 0$, are given by

$$x = \frac{-b \pm \sqrt{(b^2 - 4ac)}}{2a}$$

Safe substituting

✓ Equation is in the form $ax^2 + bx + c = 0$

✓ Write down your values of a, b and c before you substitute.

✓ Use brackets when you are substituting negative numbers.

✓ Show what you have substituted in the formula.

✓ Simplify what is under the square root and write this down.

✓ The $\pm$ symbol means you need to do two calculations.

Worked example

Aiming higher

Solve $5x^2 + x + 11 = 14$

Give your solutions correct to 3 significant figures.

Show your working clearly. **(3 marks)**

$5x^2 + x - 3 = 0$

$a = 5, b = 1, c = -3$

$x = \dfrac{-1 \pm \sqrt{1^2 - 4 \times 5 \times (-3)}}{2 \times 5}$

$= \dfrac{-1 + \sqrt{61}}{10}$ or $\dfrac{-1 - \sqrt{61}}{10}$

$= 0.681024\ldots$ or $-0.881024\ldots$

$= 0.681$ or -0.881 (to 3 s.f.)

You are asked to find 'solution**s**'. This tells you that you are solving a quadratic equation.

You must give your answer 'correct to 3 significant figures'. This tells you that you need to use the quadratic formula.

Write down at least five figures after the decimal point on the calculator display before giving your final answer. You might need to use the [S⇔D] button on your calculator to get your answer as a decimal.

How many solutions?

A quadratic equation can have two solutions, one solution or no solutions.

You can use $b^2 - 4ac$ (the part under the square root) to work out how many solutions a quadratic equation has.

You can't calculate the square root of a negative number.

If $b^2 - 4ac$ is negative, there are no solutions.

If $b^2 - 4ac = 0$ there is only one solution.

± 0 appears in the formula, so you get the same answer whether you use + or −

If $b^2 - 4ac > 0$ there are two different solutions.

Now try this

1 Solve, giving your solutions correct to 2 decimal places.

   (a) $7x^2 + 3x - 6 = 0$ **(2 marks)**

   (b) $m^2 + 50m = 6000$ **(3 marks)**

2 Solve this quadratic equation.
$2x(x - 5) = 3x - 1$
Give your solutions correct to 3 significant figures. **(3 marks)**

Worked solution video

Aiming higher

Completing the square

If a quadratic expression is written in the form $(x + p)^2 + q$ it is in **completed square** form. You can solve quadratic equations which don't have integer answers by completing the square.

Useful identities

If you learn these two identities you can save time when you are completing the square.

 1 $x^2 + 2bx + c \equiv (x + b)^2 - b^2 + c$

2 $x^2 - 2bx + c \equiv (x - b)^2 - b^2 + c$

There is more about identities on page 62.

Positive and negative roots

Remember that any positive number has **two** square roots: one **positive** and one **negative**.

If you 'square root' both sides of an equation you need to use ± (plus-or-minus) to show that there are two square roots.

$$x^2 = 4 \qquad (x + 4)^2 = 3$$
$$x = \pm 2 \qquad (x + 4) = \pm\sqrt{3}$$
$$x = -4 \pm \sqrt{3}$$

Worked example
Aiming higher

(a) Find values of p and q such that
$$x^2 + 6x - 20 \equiv (x + p)^2 + q \qquad \textbf{(2 marks)}$$

$x^2 + 6x - 20 = (x + 3)^2 - 3^2 - 20$
$\qquad\qquad = (x + 3)^2 - 9 - 20$
$\qquad\qquad = (x + 3)^2 - 29$
$\qquad\qquad p = 3 \text{ and } q = -29$

(b) Hence, or otherwise, solve the equation
$$x^2 + 6x - 20 = 0$$
Give your answer in surd form. **(2 marks)**

$(x + 3)^2 - 29 = 0 \qquad (+29)$
$\qquad (x + 3)^2 = 29 \qquad (\sqrt{\ })$
$\qquad\quad x + 3 = \pm\sqrt{29} \quad (-3)$
$\qquad\qquad\quad x = -3 \pm\sqrt{29}$

Compare the expression with the identities for completing the square.
$$x^2 + 6x - 20 \equiv (x + p)^2 + q$$
$$x^2 + 2bx + c \equiv (x + b)^2 - b^2 + c$$
$2b = 6$ so $b = 3$
$c = -20$
Substitute these values into the identity and simplify to find p and q.

Use your answer to part (a) to write the expression in completed square form. The unknown only appears once so you can solve it using **inverse operations**.
Remember to use the ± symbol when you take square roots of both sides. The two solutions are $x = -3 + \sqrt{29}$ and $x = -3 - \sqrt{29}$

Now try this

Aiming higher

1 Solve, giving your answers in surd form
(a) $x^2 + 10x + 12 = 0$ **(4 marks)**
(b) $y^2 - 6y - 15 = 0$ **(4 marks)**

2 Write $2x^2 + 20x + 7$ in the form
$a(x + p)^2 + q$ **(3 marks)**

Worked solution video

You need to be able to do these **without** a calculator.

You can start by writing the expression as $2(x^2 + 10x) + 7$, then writing $x^2 + 10x$ in completed square form.

Simultaneous equations 1

Simultaneous equations have two unknowns. You need to find the values for the two unknowns that make **both** equations true.

Algebraic solution

1. Number each equation.

2. If necessary, multiply the equations so that the coefficients of one unknown are the same.

3. Add or subtract the equations to **eliminate** that unknown.

4. Once one unknown is found use substitution to find the other.

5. Check the answer by substituting both values into the other equation.

$$3x + y = 20 \quad ①$$
$$x + 4y = 14 \quad ②$$

$$12x + 4y = 80 \quad ① \times 4$$
$$- (x + 4y = 14) \quad -②$$
$$\overline{}$$
$$11x = 66$$
$$x = 6$$

Substitute $x = 6$ into ①:
$$3(6) + y = 20$$
$$18 + y = 20$$
$$y = 2$$
Solution is $x = 6$, $y = 2$
Check it: $x + 4y = 6 + 4(2) = 14$ ✓

Worked example

Solve the simultaneous equations
$$6x + 2y = -3 \quad ①$$
$$4x - 3y = 11 \quad ② \qquad \textbf{(4 marks)}$$

$$18x + 6y = -9 \quad ① \times 3$$
$$+ \ 8x - 6y = 22 \quad ② \times 2$$
$$\overline{}$$
$$26x = 13$$
$$x = \tfrac{1}{2}$$

Substitute $x = \tfrac{1}{2}$ into ①:
$$6\left(\tfrac{1}{2}\right) + 2y = -3$$
$$3 + 2y = -3$$
$$2y = -6$$
$$y = -3$$

Easier eliminations

You can save time by choosing the right unknown to eliminate. Look for one of these:

1   If an unknown appears **on its own** in one equation, you only need to multiply one equation to eliminate that unknown.

2   If an unknown has **different signs** in the two equations, you can eliminate by **adding**.

 Multiply both equations by a whole number to make the coefficients of y the same.

 You can give your final answer as $x = \tfrac{1}{2}$, $y = -3$, or as $\left(\tfrac{1}{2}, -3\right)$.

Graphical solution

You can solve these simultaneous equations by drawing a graph.
$$x - y = 1 \qquad x + 2y = 4$$
The coordinates of the point of intersection give the solution to the simultaneous equations.
The solution is $x = 2$, $y = 1$

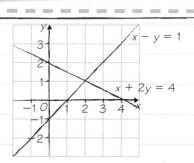

Now try this

1 Solve the simultaneous equations
$$3x - 2y = 12$$
$$x + 4y = 11 \qquad \textbf{(3 marks)}$$

 Draw a coordinate grid from -3 to 5 in both directions.

2 By drawing two suitable straight lines on a coordinate grid, solve these simultaneous equations.
$$3y + 2x = 6$$
$$y = 2x - 2 \qquad \textbf{(4 marks)}$$

Worked solution video

Simultaneous equations 2

If a pair of simultaneous equations involves an x^2 or a y^2 term, you need to solve them using **substitution**. Remember to **number** the equations to keep track of your working.

Rearrange the linear equation to make y the subject.

$$y = x^2 - 2x - 7 \quad ①$$
$$x - y = -3 \quad ②$$
$$y = x + 3 \quad ③$$
$$x + 3 = x^2 - 2x - 7 \quad \text{Substitute ③ into ①.}$$

Each solution for x has a corresponding value of y. Substitute into ③ to find the two solutions.

$$0 = x^2 - 3x - 10$$
$$0 = (x - 5)(x + 2)$$
$$x = 5 \text{ or } x = -2$$

The solutions are $x = 5$, $y = 8$ and $x = -2$, $y = 1$

Worked example

Aiming higher

Solve the simultaneous equations

$x - 2y = 1 \quad ①$

$x^2 + y^2 = 13 \quad ②$ **(5 marks)**

$x = 1 + 2y \quad ③$

Substitute ③ into ②:

$(1 + 2y)^2 + y^2 = 13$

$1 + 4y + 4y^2 + y^2 = 13$

$5y^2 + 4y - 12 = 0$

$(5y - 6)(y + 2) = 0$

$y = \frac{6}{5}$ or $y = -2$

$x = 1 + 2\left(\frac{6}{5}\right)$ $x = 1 + 2(-2)$

$\quad = \frac{17}{5}$ $\quad = -3$

Solutions: $x = \frac{17}{5}$, $y = \frac{6}{5}$ and

$\qquad x = -3$, $y = -2$

This equation has **two pairs** of solutions. Each solution is an x-value **and** a y-value. You need to find four values in total, and pair them up correctly.

You can substitute for x or y. It is easier to substitute for x because there will be no fractions.

Use brackets to make sure that the whole expression is squared.

Rearrange the quadratic equation for y into the form $ay^2 + by + c = 0$

Factorise the left-hand side to find two solutions for y.

Substitute each value of y into one of the original equations to find the corresponding values of x.

Thinking graphically

The solutions to simultaneous equations correspond to the points where the graphs of each equation **intersect**. Because an equation involving x^2 or y^2 represents a **curve**, there can be two points of intersection. Each point has an x-value and a y-value. You can write the solutions using coordinates.

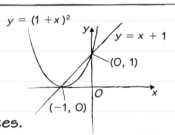

Now try this

Aiming higher

1 Solve the simultaneous equations
$x - 2y = 3$
$x^2 + 2y^2 = 27$ **(5 marks)**

2 L is the straight line with equation $3y - 2x = 5$
C is the curve with equation $y = x^2 - 1$
The line intersects the curve at two points.
Find the coordinates of both points.   **(5 marks)**

Inequalities

An inequality tells you when one value or expression is bigger or smaller than another value. You can represent **inequalities** on a number line.

$x > -1$

-3 -2 -1 0 1 2 3 4

Use an **open** circle for $>$ and $<$

The open circle shows that -1 is **not** included.

$x \leqslant 3$

-3 -2 -1 0 1 2 3 4

Use a **closed** circle for $\geqslant$ and $\leqslant$

The closed circle shows that 3 **is** included.

Solving inequalities

You can solve an inequality in exactly the same way as you solve an equation.

$x - 3 \leqslant 12$ $(+ 3)$

    $x \leqslant 15$

The solution has the letter on its own on one side of the inequality and a number on the other side.

Golden rule

If you **multiply** or **divide** both sides of an inequality by a **negative** number you have to **reverse** the **inequality** sign.

$6 - 5x > 10$ $(- 6)$

   $-5x > 4$ $(\div -5)$

     $x < \dfrac{-4}{5}$

You have divided by a negative number so you have to reverse the inequality sign.

Worked example

Solve $8x - 7 > 3x + 3$ **(2 marks)**

          $(+ 7)$

$8x > 3x + 10$ $(- 3x)$

$5x > 10$ $(\div 5)$

  $x > 2$

This is an **inequality** and not an equation. So don't use an '=' sign in your answer. The solution has the letter on its own on one side and a number on the other.

You could also use **set notation** to write your answer. You would write $\{x : x > 2\}$. This means 'the set of all values of x such that x is greater than 2'.

Integer solutions

You might need to write down all the integer solutions that **satisfy** an inequality.

Integers are positive or negative whole numbers, including 0.

$-3 \leqslant x < 2$

-4 -3 -2 -1 0 1 2 3 4 5

This shows that x is between -3 and 2. It can equal -3 but cannot equal 2.

The integer solutions that satisfy this inequality are $-3, -2, -1, 0$ and 1.

Now try this

1. Solve the inequality
   (a) $3n + 8 > 2$ **(2 marks)**
   (b) $2(n - 5) \geqslant n + 12$ **(2 marks)**

2. Find the integer value of x that satisfies both these inequalities.
   $x - 5 > 2$ $3x - 10 < 17$ **(3 marks)**

3. Find all the integer values of n that satisfy both these inequalities.
   $n + 5 > 2$ $4 - 2n \geqslant 1$ **(3 marks)**

4. Solve the inequality
   $\dfrac{x + 1}{3} < \dfrac{1 - 2x}{2}$ **(3 marks)**

Trigonometric graphs

You need to recognise and be able to sketch the graphs of the **trigonometric functions** sin, cos and tan. Revise the trigonometric functions on pages 77–78.

1 $y = \sin x$ and $y = \cos x$

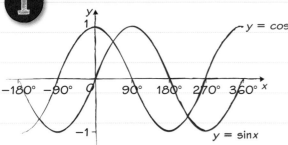

- $y = \sin x$ is the **same shape** as $y = \cos x$
- $y = \sin x$ is **translated** 90° to the right
- $y = \cos x$ is **symmetrical** about the y-axis
- $y = \sin x$ is **symmetrical** about the line $x = 90°$.

2 $y = \tan x$

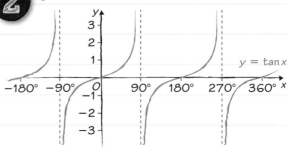

- $y = \tan x$ **repeats** every 180°
- There are **asymptotes** at −90°, 90°, 270°,...
- The graph gets closer to these asymptotes but never reaches them.

Worked example   *Aiming higher*

(a) Sketch the graph of $y = \tan x°$ for values of x from 0 to 360.   **(3 marks)**

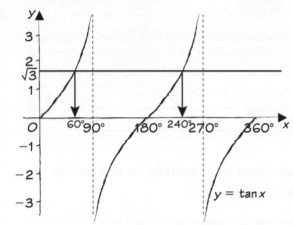

(b) $\tan 60° = \sqrt{3}$
Write down one other value of x that satisfies $\tan x° = \sqrt{3}$ **(2 marks)**
$60° + 180° = 240°$

Sketching a trig graph

✓ Label the x-axis in multiples of 90°

✓ For sin and cos label the y-axis from −1 to 1

✓ For tan label the y-axis from −3 to 3

✓ Mark some values that you know on your graph

For example $\cos 0° = 1$ and $\cos 90° = 0$.

You can use the graph to find trig values for any angle. The graph of $y = \tan x$ **repeats** every 180°, so the value of $\tan x$ will be the same as the value of $\tan(180° + -x)$. You can check this result using a calculator.

Now try this   *Aiming higher*

(a) Sketch the graph of $y = \cos x°$ for values of x from 0 to 360.   **(3 marks)**

(b) $\cos 45° = \dfrac{1}{\sqrt{2}}$. Write down one other value of x that satisfies $\cos x° = \dfrac{1}{\sqrt{2}}$.   **(2 marks)**

Transforming graphs 1

You can change the equation of a graph to translate it, stretch it or reflect it.
In the exam you might have to use functions to describe these transformations.

Function	$y = f(x) + a$	$y = f(x + a)$	$y = af(x)$
Transformation of graph	Translation $\begin{pmatrix} 0 \\ a \end{pmatrix}$	Translation $\begin{pmatrix} -a \\ 0 \end{pmatrix}$	Stretch in the vertical direction, scale factor a
Useful to know	$f(x) + a \rightarrow$ move UP a units $f(x) - a \rightarrow$ move DOWN a units	$f(x + a) \rightarrow$ move LEFT a units $f(x - a) \rightarrow$ move RIGHT a units	x-values stay the same
Example	$y = f(x) + 3$ $y = f(x)$	$y = f(x)$ $y = f(x + 5)$	$y = 3f(x)$ $y = f(x)$

Function	$y = f(ax)$	$y = -f(x)$	$y = f(-x)$
Transformation of graph	Stretch in the horizontal direction, scale factor $\frac{1}{a}$	Reflection in the x-axis	Reflection in the y-axis
Useful to know	y-values stay the same	'$-$' outside the bracket	'$-$' inside the bracket
Example	$y = f(2x)$ $y = f(x)$	$y = f(x)$ $y = -f(x)$	$y = f(-x)$ $y = f(x)$

Worked example
Aiming higher

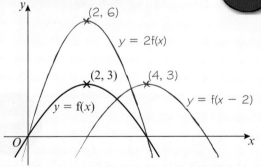

The curve $y = f(x)$ has a vertex at $(2, 3)$.
Write down the coordinates of the vertex of the curve with equation

(a) $y = f(x - 2)$ $(4, 3)$
(b) $y = 2f(x)$ $(2, 6)$

$y = f(x - 2)$ is a translation 2 units right along the x-axis.
$y = 2f(x)$ is a stretch in the vertical direction, scale factor 2.

Now try this
Aiming higher

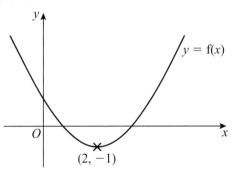

The curve $y = f(x)$ has a minimum point at $(2, -1)$.

(a) Write down the coordinates of the minimum point of the curve with equation
   (i) $y = f(x + 2)$
   (ii) $y = 3f(x)$
   (iii) $y = f(2x)$ **(3 marks)**

The curve $y = f(x)$ is reflected in the y-axis.

(b) Find the equation of the curve following this transformation.   **(1 mark)**

Transforming graphs 2

You need to be able to convert between **function notation** and equations of graphs.
This table shows some transformations that may come up in your exam.

Original function	$y = 2x + 3$	$y = \sin x°$	$y = x^2 - 2x + 1$	$y = x^2$
Transformation	$f(x) \rightarrow f(x) + 2$	$f(x) \rightarrow f(x - 30)$	$f(x) \rightarrow 2f(x)$	$f(x) \rightarrow f(3x)$
Which means...	movement UP by 2 units	movement RIGHT by 30°	stretch in vertical direction, scale factor 2	stretch in horizontal direction, scale factor $\frac{1}{3}$
New function	$y = 2x + 5$	$y = \sin(x - 30)°$	$y = 2x^2 - 4x + 2$	$y = 9x^2$

Graphs of sine and cosine

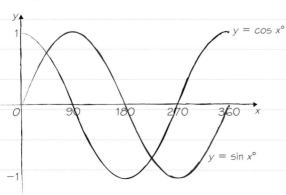

The graph of $y = \cos x°$ is identical to the graph of $y = \sin x°$ except that it has been moved to the left by 90°.

Write down the transformations using function notation.
(a) Stretch in the vertical direction with scale factor $\frac{1}{2}$.
(b) Stretch in the horizontal direction with scale factor 2.

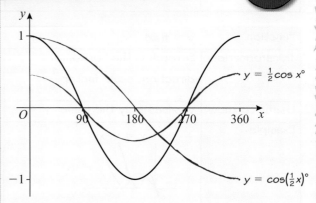

The diagram shows a sketch of the curve $y = \cos x°$ for $0 \leqslant x \leqslant 360$

On the same diagram sketch the curve with equation

(a) $y = \frac{1}{2} \cos x°$ $y = \frac{1}{2}f(x)$
(b) $y = \cos\left(\frac{1}{2}x\right)°$ $y = f\left(\frac{1}{2}x\right)$

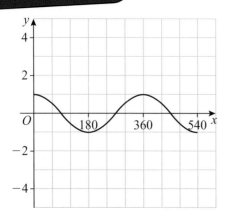

The grid shows the graph of $y = \cos x°$ for values of x from 0 to 540.

On the grid, sketch the graph of $y = 3\cos(2x)°$ for values of x from 0 to 540.   **(2 marks)**

Inequalities on graphs

You can show the points that satisfy inequalities involving x and y on a graph.

For example, follow these steps to shade the region R that satisfies the inequalities

$$x \geqslant 2 \qquad y > x \qquad x + y \geqslant 6$$

Always work on one inequality at a time.

 1

$x \geqslant 2$
Draw the graph of $x = 2$ with a solid line.
Use a small arrow to show which side of
the line you want.

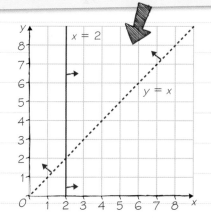 **2**

$y > x$
Draw the graph of $y = x$ with a
dotted line.
Show which side of the line you want.

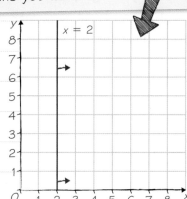

 3

$x + y \geqslant 6$
Draw the graph of $x + y = 6$ with a
solid line. Use a table of values.

x	0	3	6
y	6	3	0

Show which side of the line you want.
$x + y$ increases as you move away from
the origin.
Shade in the region and label it **R**.

 4

Check it!
Pick a point inside your shaded region.
Check that the x- and y-values for that
point satisfy **all** the inequalities.
At (4, 5) $x = 4$ and $y = 5$.
$x \geqslant 2$ ✓
$y > x$ ✓
$x + y \geqslant 6$ ✓

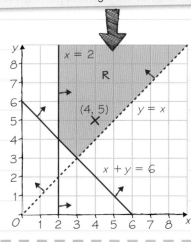

Graphical inequalities checklist

☑ $<$ and $>$ are shown by **dotted** lines.

☑ $\leqslant$ and $\geqslant$ are shown by **solid** lines.

☑ Points on a solid line **are** included in
the region.

☑ Points on a dotted line **aren't** included
in the region.

Now try this

Aiming higher

Worked solution video

Draw a coordinate grid from –2 to 8 in both directions. Shade the region
of points whose coordinates satisfy these inequalities.
$x \geqslant -1 \qquad y > 5 \qquad y \geqslant x + 3$

(4 marks)

Sketching graphs

A **sketch** of a graph shows all the major features. These usually include turning points and places where the graph crosses the axes.

Factors and roots

If a function is **factorised** you can work out the points where its graph crosses the x-axis. These are called the **roots** of the function. You can find them by working out the x-values that make each factor equal to **zero**.

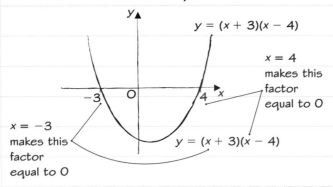

$y = (x + 3)(x - 4)$

$x = 4$ makes this factor equal to 0

$x = -3$ makes this factor equal to 0

$y = (x + 3)(x - 4)$

Graph sketching checklist

☑ Your sketch should be **clear** and **neat**.

☑ Use a **ruler** for any **straight** lines.

☑ Label your **axes** and the **origin**.

☑ Label your graph with its **equation**.

☑ Label any **turning points** if you are asked to find them.

There is more about finding turning points of quadratic graphs on page 44.

☑ Label any points where the graph crosses the axes.

The question will usually state exactly what points you need to show.

Sketching cubics – two to watch

1 If one factor is x then the curve will pass through the origin.

2 If one factor is **squared** then the curve will **just touch** the x-axis at the corresponding point.

Worked example

Aiming higher

The factor $(x + 3)$ is **squared** (or **repeated**). This means that the curve **just touches** the x-axis at the point that makes this factor equal to zero. Be careful with your x-values. The factor $(x + 3)$ is zero when $x = -3$ and the factor $(x - 2)$ is zero when $x = 2$.

Don't forget to work out the point where the curve intercepts the y-axis. You do this by setting $x = 0$ in the function:
$y = (0 - 2)(0 + 3)^2 = -2 \times 9 = -18$

$f(x) = (x - 2)(x + 3)^2$
Sketch the graph of $y = f(x)$, showing the coordinates of any intercepts with the coordinate axes. **(3 marks)**

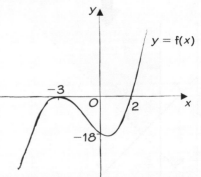

$y = f(x)$

Now try this

Aiming higher

Sketch graphs of the following equations, showing the coordinates of any intercepts with the coordinate axes.

(a) $y = x^2 - 7x + 12$ **(2 marks)**

(b) $y = x(x^2 + 4x - 12)$ **(3 marks)**

Using quadratic graphs

You can use graphs to solve quadratic equations. You will need to look for the x-values of the points where a quadratic graph intersects with a straight line.

This grid shows one quadratic graph and two straight-line graphs.

The x-values at A and B are the solutions of the quadratic equation

$$x^2 - 4x = 12$$

or

$$x^2 - 4x - 12 = 0$$

The x-values at C and D are the solutions of the quadratic equation

$$x^2 - 4x = x + 3$$

or

$$x^2 - 5x - 3 = 0$$

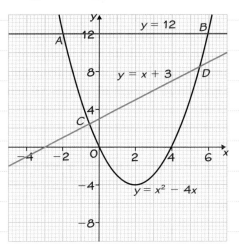

Worked example

This is a graph of $y = 2x^2 + 5x$

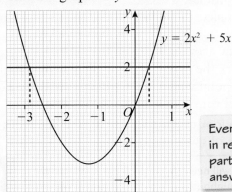

Everything in red is part of the answer.

By drawing a suitable straight line on the graph, find estimates for the solutions of the equation $2x^2 + 5x - 2 = 0$

Give your answers correct to 1 decimal place. **(3 marks)**

$$2x^2 + 5x - 2 = 0 \qquad (+ 2)$$
$$2x^2 + 5x = 2$$
$$x = 0.4, x = -2.9$$

You can solve the **quadratic equation** $2x^2 + 5x - 2 = 0$ by finding where the graph $y = 2x^2 + 5x$ crosses the straight line $y = 2$

Draw the line $y = 2$ on the graph.

The solutions are the x-values at the points of intersection.

A suitable line

To find a **suitable** straight line, rearrange the quadratic equation so that the left-hand side matches the equation of the quadratic graph. The right-hand side will tell you which line to draw. Here is another example:

Graph given: $y = x^2 - 6x + 4$

Equation to solve: $x^2 - 5x + 1 = 0$

Rearrange equation: $x^2 - 6x + 4 = -x + 3$

Line to draw: $y = -x + 3$

Now try this

Aiming higher

(a) Complete the table of values for $y = x^2 + 3x - 3$

x	−5	−4	−3	−2	−1	0	1	2
y	7	1			−3	1		7

(2 marks)

(b) On a grid with $-5 \leqslant x \leqslant 5$ and $-8 \leqslant y \leqslant 8$, draw the graph of $y = x^2 + 3x - 3$ **(2 marks)**

(c) Find estimates for the solutions of $x^2 + 3x - 3 = 0$ **(1 mark)**

(d) By drawing a suitable straight line, work out the solutions of the equation $x^2 + 2x - 4 = 0$ **(3 marks)**

Turning points

You can find the turning point of a quadratic graph by writing the function in **completed square** form. Make sure you are confident with completing the square before tackling this page. Have a look at page 34 for a reminder.

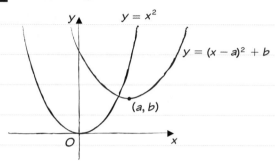

$y = x^2$

$y = (x - a)^2 + b$

(a, b)

Golden rule

The graph of
$y = (x - a)^2 + b$
has a turning point
at (a, b)

LEARN IT!

The graph of $y = (x - a)^2 + b$ is a translation of the graph of $y = x^2$ by the vector $\binom{a}{b}$. There is more on transformations of graphs on page 39.

Worked example
Aiming higher

(a) $f(x) = x^2 + 3x + 5$

Sketch the graph of $y = f(x)$, showing the coordinates of the turning point and the coordinates of any intercepts with the coordinate axes. **(4 marks)**

$x^2 + 3x + 5 = (x + 1.5)^2 - 1.5^2 + 5$
$= (x + 1.5)^2 + 2.75$

so turning point at $(-1.5, 2.75)$
When $x = 0$, $f(x) = 5$ so graph intercepts y-axis at $(0, 5)$

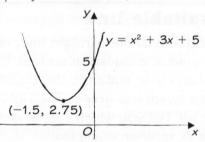

$y = x^2 + 3x + 5$

5

$(-1.5, 2.75)$

(b) Hence, or otherwise, determine whether $f(x)$ has any real roots. Give reasons for your answer. **(2 marks)**

$y = f(x)$ does not intercept x-axis, so $f(x)$ has no real roots.

Problem solved!

If a question asks you to **sketch** the graph of a quadratic function, you should use **algebra** to find the turning point and any intercepts with the coordinate axes.

1. **Complete the square** to find the turning point.
2. Set $x = 0$ to work out the y-intercept.
3. Draw axes with a ruler, then sketch the graph.
4. Mark the coordinates of the turning point and the point where the graph crosses the y-axis, and label your graph.

Have a look at page 29 for a reminder about the shapes of quadratic graphs.

You will need to use problem-solving skills throughout your exam – **be prepared!**

Functions and roots

The roots of a function $f(x)$ are the values of x for which $f(x) = 0$. This means that the roots of the function are the x-values at the points where $y = f(x)$ crosses the x-axis.

Now try this
Aiming higher

The diagram shows a sketch of the curve with equation $y = x^2 - 4x - 3$

(a) Write down the coordinates of the point A, where the curve crosses the y-axis. **(1 mark)**

(b) By completing the square for $x^2 - 4x - 3$ find the coordinates of the turning point B. **(3 marks)**

Worked solution video

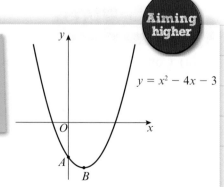

$y = x^2 - 4x - 3$

A
B

Quadratic inequalities

On page 37 you revised **linear inequalities**. In your exam, you might need to solve a **quadratic inequality**, which involves an x^2 term.

Using a sketch

This graph shows a **sketch** of the **curve** $y = x^2$, and the **straight line** $y = 36$

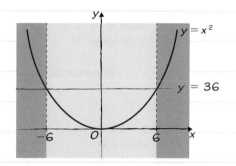

☐   For these values of x, the curve is **below** the line, so $x^2 < 36$

The solutions of $x^2 < 36$ are $-6 < x < 6$

☐   For these values of x, the curve is **above** the line, so $x^2 > 36$

The solutions of $x^2 > 36$ are $x < -6$ or $x > 6$

Worked example   *Aiming higher*

Solve the inequality $m^2 \leqslant 25$ **(2 marks)**

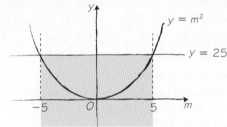

$-5 \leqslant m \leqslant 5$

Draw a sketch of $y = m^2$ and $y = 25$
$\sqrt{25} = 5$ so the graphs intersect at
$m = 5$ and $m = -5$
You want m^2 to be **less than or equal to** 25, so you need to consider the values of m between -5 and 5. Make sure you use the same type of inequality signs as that given in the question.

Check it!

Choose a value in your solution range and check it satisfies the original inequality:
$(-3)^2 = 9 \leqslant 25$ ✓

Worked example   *Aiming higher*

Solve the inequality $2x^2 - 5x - 3 > 0$
Represent your solution on this number line.

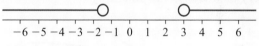

(3 marks)

$(2x + 1)(x - 3) > 0$

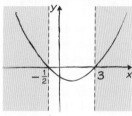

$x < -\frac{1}{2}$ or $x > 3$

Factorise the left-hand side then sketch the graph. The inequality is $>$ so you are interested in the x-values where the graph is **above** the horizontal axis.

Now try this

1 Solve the inequality $p^2 < 4$
Aiming higher
Represent your solution on this number line.

$-6\ -5\ -4\ -3\ -2\ -1\ \ 0\ \ 1\ \ 2\ \ 3\ \ 4\ \ 5\ \ 6$

(2 marks)

2 Find the set of values of x for which
Aiming higher
   $x^2 - 5x - 14 < 0$ **(4 marks)**

Start by sketching the graph of $y = x^2 - 5x - 14$. You are interested in the values of x where the graph is **below** the horizontal axis.

Gradients of curves

You can estimate the gradient of a curve at a given point by drawing a **tangent** to the curve at that point. This distance–time graph shows a runner accelerating at the start of a race.

On a distance–time graph the gradient tells you the speed. There is more about this on page 31.

This straight line is the **tangent** to the curve at $t = 3$.

Draw a large triangle to make your estimate as accurate as possible.

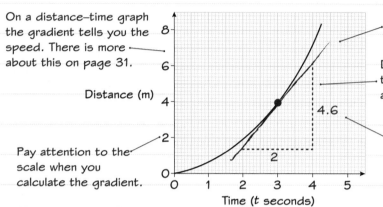

Pay attention to the scale when you calculate the gradient.

$\dfrac{4.6}{2} = 2.3$

so after 3 seconds the runner was travelling at approximately 2.3 m/s.

Worked example
Aiming higher

This graph shows the voltage across a phone battery as it charges from empty.

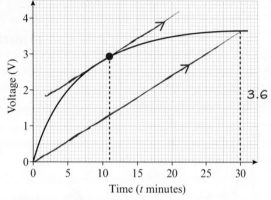

(a) Work out the average rate of increase of voltage between $t = 0$ and $t = 30$. **(2 marks)**

$\dfrac{3.6}{30} = 0.12$ V/min

Amir wants to stop charging his phone when the rate of increase of voltage drops to this average level.

(b) After how long should Amir stop charging his phone? You must show how you got your answer. **(2 marks)**

11 minutes

Problem solved!

(a) To work out the average rate of change between $t = 0$ and $t = 30$ draw a straight line between these points on the graph and find its gradient. This is the same as working out

$\dfrac{\text{change in voltage}}{\text{change in time}}$

Look at the axis labels to work out the correct units.

(b) You need to find the point **on the curve** with the same gradient. That means you need to find a **tangent** to the curve that is parallel to the first line. Slide a transparent ruler across the graph until it just touches the curve. You can show how you got your answer by drawing the tangent to the curve at this point.

> You will need to use problem-solving skills throughout your exam – **be prepared!**

Now try this
Aiming higher

A container is filling with water. This graph shows the depth of water in the container.

(a) Work out the average rate of increase of depth between $t = 0$ and $t = 40$. **(2 marks)**

(b) Use the graph to estimate the rate of increase of depth of water at $t = 10$. **(2 marks)**

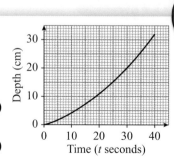

Proportion and graphs

You can use the symbol $\propto$ to show proportion. You can show quantities that are **directly proportional** or **inversely proportional** on a graph.

Direct proportion facts

If y is directly proportional to x:

✓   you can write $y \propto x$

✓   you can write an equation $y = kx$ where k is a number

✓   the graph of x against y is a **straight line** passing **through the origin**.

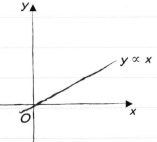

Inverse proportion facts

If y is inversely proportional to x:

✓   you can write $y \propto \dfrac{1}{x}$ y is directly proportional to the **reciprocal** of x

✓   you can write an equation $y = \dfrac{k}{x}$ where k is a number

✓   the graph of x against y looks like a **reciprocal graph**.

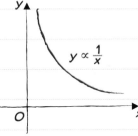

---

p and q are inversely proportional. Circle the equation that could describe the relationship between p and q.

$p = 2q$ $p = q + 5$ $p = \dfrac{q}{10}$ $\boxed{p = \dfrac{2}{q}}$ **(1 mark)**

> An equation for **inverse** proportionality looks like $y = \dfrac{k}{x}$ where k is a number.

There are other ways to answer this question. The method used here is a bit like using **equivalent ratios**:

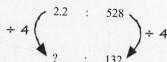

Check it!
Your answer should make sense. If the quantities are in **direct proportion** then when one decreases the other should decrease as well. ✓

Worked example

Electrical appliances in your home follow this rule:
Power (watts) $\propto$ Current (amps)

An electric drill uses 2.2 A and has a power of 528 W.

Calculate the current used by a television with a power of 132 W.   **(2 marks)**

$528 \div 132 = 4$
$2.2 \div 4 = 0.55$
The television uses 0.55 A

This graph can be used to convert between inches and centimetres.

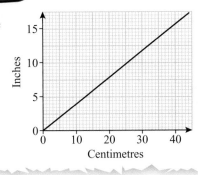

(a) Use the graph to convert 10 cm to inches.   **(1 mark)**

(b) Use the graph to convert 12.5 inches to cm.   **(1 mark)**

(c) What evidence is there from the graph to show that inches are directly proportional to centimetres?   **(2 marks)**

Proportionality formulae

You can answer some questions involving proportion by constructing a **formula**. On this page you can revise finding formulae for the two basic proportionality relationships.

1 Direct proportion

These all mean the same thing:

- y is directly proportional to x
- y varies directly with x
- $y \propto x$
- $y = kx$ — k is called the **constant of proportionality**.

2 Inverse proportion

These all mean the same thing:

- y is inversely proportional to x
- y varies inversely with x
- $y \propto \dfrac{1}{x}$
- $y = \dfrac{k}{x}$

Worked example

Winnie drops a stone down a well. The speed of the stone, v m/s, is directly proportional to the time, t seconds, since she dropped it.
After 0.5 seconds the stone is travelling at 4.9 m/s.

(a) Find a formula for v in terms of t. **(3 marks)**

$v = kt$
$4.9 = k(0.5)$ $\quad$ $(\div 0.5)$
$k = 9.8$
$v = 9.8t$

(b) Calculate the speed of the stone after 1.2 seconds. **(1 mark)**

$v = 9.8(1.2) = 11.76$ m/s

You can find a proportionality formula if you know the type of proportionality and are given two corresponding values for the variables. Follow these steps:

1. Write down the formula using k for the constant of proportionality.
2. Substitute the values of v and t you are given.
3. Solve the equation to find the value of k.
4. Write down the formula putting in the value of k.
5. Once you have written your formula, you can use it to find the value of one variable if you know the value of the other.

Checking for proportionality

You can use a graph to check whether two quantities are directly proportional.

P	5	10	15
Q	1.2	1.5	1.8

The graph doesn't go through the origin so P and Q are not directly proportional.

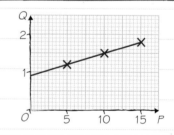

Now try this

1 x is directly proportional to y.
When $x = 36$, $y = 5$

(a) Find a formula for x in terms of y. **(3 marks)**

(b) Calculate the value of x when $y = 32$ **(1 mark)**

(c) Calculate the value of y when $x = 9$ **(1 mark)**

Worked solution video

2 A machine makes solid plastic cylinders of different heights and radii.

The height h cm of a plastic cylinder is inversely proportional to its radius r cm.

A plastic cylinder of height 6 cm has a radius of 4 cm.

Work out the height of a plastic cylinder with a radius of 3 cm. **(4 marks)**

Harder relationships

You can use proportion to describe harder relationships between variables.
Here are the relationships and graphs you should be familiar with.

Proportionality in words	Using $\propto$	Formula
y is directly proportional to the square of x	$y \propto x^2$	$y = kx^2$
y is directly proportional to the cube of x	$y \propto x^3$	$y = kx^3$
y is directly proportional to the square root of x	$y \propto \sqrt{x}$	$y = k\sqrt{x}$
y is inversely proportional to the square of x	$y \propto \dfrac{1}{x^2}$	$y = \dfrac{k}{x^2}$

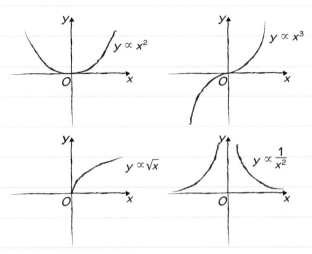

Worked example

Aiming higher

q varies inversely with the square of t.

(a) What happens to q if t is doubled? **(1 mark)**

It is divided by 4.

(b) When $t = 4$, $q = 8.5$
    Calculate the value of q when $t = 5$
    (4 marks)

$q \propto \dfrac{1}{t^2}$

$q = \dfrac{k}{t^2}$

$8.5 = \dfrac{k}{4^2}$

$k = 8.5 \times 4^2 = 136$

$q = \dfrac{136}{t^2}$

When $t = 5$: $q = \dfrac{136}{5^2} = 5.44$

'q varies inversely with the square of t' means 'q is inversely proportional to t^2'.

$q = \dfrac{k}{t^2}$ so if t is doubled the new value of q is:

$\dfrac{k}{(2t)^2} = \dfrac{k}{4t^2} = \dfrac{1}{4}q$

Always...

1. Write down the statement of proportionality and then the formula.
2. Substitute the values you are given.
3. Solve the equation to find k.
4. Write down the formula using the value of k.
5. Use your formula to find any unknown values.

Now try this

Aiming higher

1 A stone is dropped off a cliff. It takes t seconds to fall a distance d metres.

t is directly proportional to $\sqrt{d}$

When $t = 4.6$, $d = 25$

(a) Express t in terms of d. **(3 marks)**

(b) Find the length of time the stone takes to fall 42.25 m. **(2 marks)**

Aiming higher

2 The speed v (km/s) of a satellite is inversely proportional to the square root of the radius of its orbit r (km).

The Hubble space telescope orbits at a radius of 6940 km at a speed of 7.6 km/s.

Work out the speed of a satellite orbiting at a radius of 42 000 km. Give your answer to 3 significant figures. **(4 marks)**

Worked solution video

Rearranging formulae

Most formulae have one letter on its own on one side of the formula. This letter is called the **subject** of the formula.

$$E = mc^2 \qquad E \text{ is the subject of the formula.}$$

Changing the subject of a formula is like solving an equation. You have to do the same thing to both sides of the formula until you have the new letter on its own on one side.

$$E = mc^2 \qquad (\div m)$$

$$\frac{E}{m} = c^2 \qquad (\sqrt{})$$

The inverse operation to $\boxed{x^2}$ is $\boxed{\sqrt{}}$. You need to square root **everything** on both sides of the formula.

$$\sqrt{\frac{E}{m}} = c \qquad c \text{ is now the subject of the formula.}$$

Harder formulae

If the letter you need **appears twice** in the formula you need to **factorise**.

| **Group** all the terms with that letter on one side of the formula and all the other terms on the other side. | ➡ | **Factorise** so the letter only appears once. | ➡ | **Divide** by everything in the bracket to get the letter on its own. |

For a reminder about factorising have a look at page 21.

Worked example

$$N = \frac{3h + 20}{100}$$

Rearrange the formula to make h the subject.

(2 marks)

$$N = \frac{3h + 20}{100} \qquad (\times 100)$$
$$100N = 3h + 20 \qquad (- 20)$$
$$100N - 20 = 3h \qquad (\div 3)$$
$$\frac{100N - 20}{3} = h$$
$$h = \frac{100N - 20}{3}$$

Your final answer should look like $h = ...$

Worked example

 Aiming higher

Make Q the subject of the formula $P = \dfrac{Q}{Q - 100}$

(3 marks)

$$P = \frac{Q}{Q - 100} \qquad [\times (Q - 100)]$$
$$P(Q - 100) = Q \qquad \text{(multiply out brackets)}$$
$$PQ - 100P = Q \qquad (+ 100P)$$
$$PQ = Q + 100P \qquad (- Q)$$
$$PQ - Q = 100P \qquad \text{(factorise)}$$
$$Q(P - 1) = 100P \qquad [\div (P - 1)]$$
$$Q = \frac{100P}{P - 1}$$

Your final answer should look like $Q = ...$
You need to factorise to get Q on its own.

Now try this

1 Rearrange this formula to make t the subject:
   $$4p = 3t - 1 \qquad \textbf{(2 marks)}$$

2 Make w the subject of $m = \sqrt{5w + 7}$
   (2 marks)

 Aiming higher

3 Rearrange this formula to make P the subject:
   $$Q = \sqrt{\frac{100 - 5P}{P}} \qquad \textbf{(4 marks)}$$

 Worked solution video

Sequences and series

An **arithmetic series** is a sum of the terms in an arithmetic sequence. You can use formulae to solve problems involving arithmetic sequences and series. You use a to represent the first term and d to represent the common difference.

$$\overset{+3}{\frown}\ \overset{+3}{\frown}\ \overset{+3}{\frown}\ \overset{+3}{\frown}$$
$$11 + 14 + 17 + 20 + 23 + \ldots$$
$$a = 11, d = 3$$

nth term

The nth term of an arithmetic sequence or series with first term a and common difference d is

$a + (n - 1)d$

You need to learn this – it is not on the formulae sheet.

Sum to n terms

The **sum** of the first n terms of an arithmetic series with first term a and common difference d is $S_n = \dfrac{n}{2}[2a + (n - 1)d]$

This formula appears on the formulae sheet.

Worked example
Aiming higher

The first term of an arithmetic sequence is a and the common difference is d.

The 13th term of the sequence is 8 and the 16th term of the sequence is $12\frac{1}{2}$

(a) Write down two equations for a and d.
(2 marks)

$a + (13 - 1)d = 8$ so $a + 12d = 8$ ①

$a + (16 - 1)d = 12\frac{1}{2}$ so $a + 15d = 12\frac{1}{2}$ ②

(b) Find the values of a and d. **(2 marks)**

② − ①: $3d = 4\frac{1}{2}$

$d = 1\frac{1}{2}$

Substitute $d = 1\frac{1}{2}$ into ①: $a + 12\left(1\frac{1}{2}\right) = 8$

$a + 18 = 8$

$a = -10$

> You could also work out the common difference by writing down a few terms of the sequence:
>
> $$\overset{+1\frac{1}{2}}{\frown}\ \overset{+1\frac{1}{2}}{\frown}\ \overset{+1\frac{1}{2}}{\frown}$$
> $$\ldots,\quad 8,\quad ?,\quad ?,\quad 12\frac{1}{2},\ \ldots$$
>
> There are three jumps between the 13th term and the 16th term. $12\frac{1}{2} - 8 = 4\frac{1}{2}$
> So each jump is $4\frac{1}{2} \div 3 = 1\frac{1}{2}$.

> Solve the two equations simultaneously to find a and d. Number each equation to keep track of your working.

> You can work in decimals or mixed numbers. You could give your answer as $a = -10$, $d = 1.5$.

Worked example
Aiming higher

The first four terms of an arithmetic series are
$7 + 11 + 15 + 19 + \ldots$
Find the sum of the first 50 terms of this series.
(3 marks)

$a = 7$ and $d = 4$

$S_n = \dfrac{n}{2}[2a + (n - 1)d]$

$S_{50} = \dfrac{50}{2}[2 \times 7 + (50 - 1) \times 4]$

$= 25 \times (14 + 49 \times 4) = 5250$

> The notation S_n represents the sum of the first n terms, so you can write the sum of the first 50 terms as S_{50}. Copy the formula from the formulae sheet then substitute the values of a and d.

Now try this
Aiming higher

1 The first four terms of an arithmetic series are $3 + 8 + 13 + 18 + \ldots$
  Find the sum of the first 40 terms of this series. **(3 marks)**

2 The 3rd term of an arithmetic series is 21
  The 7th term of the same arithmetic series is 57
  Find the sum of the first 50 terms of this arithmetic series. **(5 marks)**

Algebraic fractions

Simplifying an algebraic fraction is just like simplifying a normal fraction.

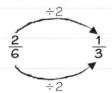

$$\frac{2}{6} \xrightarrow{\div 2} \frac{1}{3} \qquad \frac{x+1}{2x(x+1)} \xrightarrow{\div(x+1)} \frac{1}{2x}$$

You can divide the top and bottom by a number, a term or a whole expression.

Golden rule

If the top or the bottom of the fraction has **more than** one term, you will need to factorise before simplifying.

$$\frac{p^2 + 3p}{4p} = \frac{p(p + 3)}{4p} = \frac{p + 3}{4}$$

Two terms on top so factorise the top, then divide the top and bottom by p.

Operations on algebraic fractions

 To **add** or **subtract** algebraic fractions with different denominators:

1. Find a common denominator.
2. Add or subtract the numerators.
3. Simplify if possible.

The smallest common denominator isn't always the product of the two denominators.

You can use a common denominator of $4x$ to simplify this expression:

$$\frac{x+1}{2x} + \frac{3-2x}{4x}$$

$$\frac{1}{x + 4} + \frac{2}{x - 4} = \frac{x - 4}{(x + 4)(x - 4)} + \frac{2(x + 4)}{(x + 4)(x - 4)}$$

$$= \frac{x - 4 + 2x + 8}{(x + 4)(x - 4)} = \frac{3x + 4}{(x + 4)(x - 4)}$$

 To **multiply** fractions:

1. Multiply the numerators **and** multiply the denominators.
2. Simplify if possible.

$$\frac{x}{2} \times \frac{4}{x - 1} = \frac{{}^{2}\cancel{4}x}{{}_{1}\cancel{2}(x - 1)} = \frac{2x}{x - 1}$$

Don't expand brackets if you don't have to. It's much easier to simplify your fraction with the brackets in place.

 To **divide** fractions:

1. Change the second fraction to its reciprocal.
2. Change $\div$ to $\times$
3. Multiply the fractions and simplify.

$$\frac{x^2}{3} \div \frac{x}{6} = \frac{x^2}{3} \times \frac{6}{x} = \frac{{}^{2}\cancel{6}x^2}{{}_{1}\cancel{3}x} = 2x$$

To find the reciprocal of a fraction you turn it upside down.

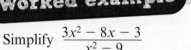 **Worked example** — Aiming higher

Simplify $\dfrac{3x^2 - 8x - 3}{x^2 - 9}$ **(3 marks)**

$$\frac{3x^2 - 8x - 3}{x^2 - 9} = \frac{(3x + 1)\cancel{(x - 3)}}{(x + 3)\cancel{(x - 3)}}$$

$$= \frac{3x + 1}{x + 3}$$

You need to factorise the top and the bottom of the fraction before you can simplify. Remember that $a^2 - b^2 = (a + b)(a - b)$

 Now try this — Aiming higher

1 Simplify fully $\dfrac{a + 1}{3a} + \dfrac{7}{6a}$ **(2 marks)**

2 Solve $\dfrac{1}{10x} - \dfrac{1}{15x} = 4$ **(3 marks)**

3 Simplify fully

(a) $\dfrac{x^2 + 4x - 12}{x^2 - 25} \div \dfrac{x + 6}{x^2 - 5x}$ **(4 marks)**

(b) $\dfrac{3m^2 - 108}{9m^3 + 54m^2}$ **(3 marks)**

Worked solution video

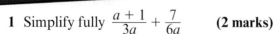 

Quadratics and fractions

You need to remove fractions before you can solve an equation.

For a reminder about solving linear equations with fractions have a look at page 23.

To remove fractions from an equation multiply everything by the lowest common multiple of the denominators.

$(2x - 3)$ and $(x + 1)$ don't have any common factors. Multiply everything by $(2x - 3)(x + 1)$

$$\frac{x}{2x - 3} + \frac{4}{x + 1} = 1$$

$$\frac{x(2x - 3)(x + 1)}{2x - 3} + \frac{4(2x - 3)(x + 1)}{x + 1} = (2x - 3)(x + 1)$$

Don't expand brackets until you have simplified the fractions.

$$x(x + 1) + 4(2x - 3) = (2x - 3)(x + 1)$$

$$x^2 + x + 8x - 12 = 2x^2 - 3x + 2x - 3$$

Multiply out brackets and collect like terms.

$$0 = x^2 - 10x + 9$$

$$= (x - 9)(x - 1)$$

The solutions are $x = 9$ and $x = 1$

Worked example

Aiming higher

Solve $\dfrac{2 + 3x}{5x + 9} = \dfrac{2}{x - 1}$ **(4 marks)**

$$\left(\frac{2 + 3x}{5x + 9}\right)(x - 1)(5x + 9) = \left(\frac{2}{x - 1}\right)(x - 1)(5x + 9)$$

$$\frac{(2 + 3x)(x - 1)(5x + 9)}{5x + 9} = \frac{2(x - 1)(5x + 9)}{x - 1}$$

$$(2 + 3x)(x - 1) = 2(5x + 9)$$

$$3x^2 - 3x + 2x - 2 = 10x + 18$$

$$3x^2 - 11x - 20 = 0$$

$$(3x + 4)(x - 5) = 0$$

$$3x + 4 = 0 \qquad\qquad x - 5 = 0$$

$$x = -\frac{4}{3} \qquad\qquad x = 5$$

Multiply everything by $(x - 1)(5x + 9)$ to remove the fractions.

If you are confident working with algebraic fractions you can jump straight to this step.

Set the two factors equal to 0 to find the two solutions.

Quadratic equations checklist

- ✓ Remove any fractions by multiplying everything by the lowest common multiple of the denominators.
- ✓ Multiply out any brackets and collect like terms.
- ✓ Rewrite in the form $ax^2 + bx + c = 0$
- ✓ Factorise the left-hand side to solve the quadratic equation.

Now try this

Multiply everything by $5(3x - 1)(2x + 1)$

Aiming higher

1 Solve $\dfrac{5}{x} + \dfrac{2}{x + 2} = 3$ **(4 marks)**

 Worked solution video

2 Solve $\dfrac{2}{3x - 1} - \dfrac{3}{2x + 1} = \dfrac{2}{5}$ **(4 marks)**

3 Solve $\dfrac{1}{2x + 3} - \dfrac{1}{x} = \dfrac{1}{20}$ **(4 marks)**

Surds 2

You might need to expand brackets involving surds. You can use the grid method or **FOIL** to expand the brackets (have a look at page 20). Here are two **golden rules** to remember when working with surds.

 You can use the rule $\sqrt{ab} = \sqrt{a} \times \sqrt{b}$ in **both directions**:

$\sqrt{20} \times \sqrt{5} = \sqrt{100} = 10$

$\sqrt{2a} \times \sqrt{18a} = \sqrt{36a^2} = \sqrt{36} \times \sqrt{a^2} = 6a$

 Whole number parts and surd parts stay **separate**:

If $24 + 6\sqrt{7} = a + b\sqrt{7}$ then you can **compare** the two expressions to get $a = 24$ and $b = 6$

Worked example

Show that $(3 + \sqrt{8})(4 + \sqrt{8}) = 20 + 14\sqrt{2}$

Show each stage of your working clearly. **(2 marks)**

$(3 + \sqrt{8})(4 + \sqrt{8}) = 12 + 3\sqrt{8} + 4\sqrt{8} + (\sqrt{8})^2$

$= 12 + 7\sqrt{8} + 8$

$= 20 + 7\sqrt{8}$

$= 20 + 7 \times 2\sqrt{2}$

$= 20 + 14\sqrt{2}$

When the question says 'Show that ...' you should start from the **left-hand side**, then simplify and rearrange until your expression matches the **right-hand side**.

Make sure you **simplify** any surds in your expression:

$\sqrt{8} = \sqrt{4 \times 2}$

$= \sqrt{4} \times \sqrt{2}$

$= 2\sqrt{2}$

There's more about this on page 8.

Rationalising denominators

You can **rationalise a denominator** in the form $a + \sqrt{b}$ by multiplying the numerator and denominator by $a - \sqrt{b}$:

$$\frac{5}{2 - \sqrt{3}} = \frac{5(2 + \sqrt{3})}{(2 - \sqrt{3})(2 + \sqrt{3})}$$

$$= \frac{10 + 5\sqrt{3}}{4 - 2\sqrt{3} + 2\sqrt{3} - 3}$$

$$= 10 + 5\sqrt{3}$$

Worked example

Aiming higher

Given that a and b are positive integers such that $\left(6 + \sqrt{a}\right)^2 = b + 12\sqrt{2}$

find the value of a and the value of b. **(3 marks)**

$(6 + \sqrt{a})^2 = (6 + \sqrt{a})(6 + \sqrt{a})$

$= 36 + 6\sqrt{a} + 6\sqrt{a} + (\sqrt{a})^2$

$= 36 + a + 12\sqrt{a}$

$= b + 12\sqrt{2}$

$a = 2$ and $b = 38$

Now try this

Aiming higher

1 Show that $= (3 - \sqrt{12})^2 = 21 - 12\sqrt{3}$
Show each stage of your working clearly. **(3 marks)**

2 $(5 + \sqrt{x})(3 + \sqrt{x}) = 18 + k\sqrt{x}$
where x is a prime number and k is a positive integer.
Find the value of x and the value of k. **(3 marks)**

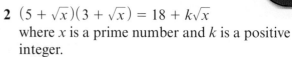

Functions

A function maps numbers in its **domain** onto numbers in its **range**. Here is an example:

f is the **name** of the function. You can use any letter, but f and g are the most common.

$$f(x) = \sqrt{x - 2} \qquad x \geqslant 2$$

This is the **domain** of the function. The function is only defined for these **input** values. The **range** of this function is f(x) $\geqslant$ 0. This tells you all the possible **output** values for the function.

x is the **input**. You say 'f of x'. You can also write $f : x \to \sqrt{x} \quad 2$ and say 'f maps x onto $\sqrt{x} - 2$'

This tells you what the function does to x.

Worked example

f is the function $f(x) = 3x + 10$

(a) Find f(−2) **(1 mark)**

$f(-2) = 3 \times (-2) + 10$
$\qquad = -6 + 10 = 4$

(b) Solve f(a) = 31 **(2 marks)**

$3a + 10 = 31 \qquad (-10)$
$\quad 3a = 21 \qquad (\div 3)$
$\qquad a = 7$

If the question doesn't ask you about the **domain** or the **range** then you don't need to worry about them.

(a) To find f(−2) you just **substitute** $x = -2$ into the expression for f(x)

(b) Substitute $x = a$ into the expression for f(x) then solve the equation to find a.

Excluding values

A function must be clearly defined. **Every value** in the **domain** can be used as an input. This means that you sometimes have to **exclude** values from the domain. Here are two cases to learn:

 You can't divide by 0

$f(x) = \dfrac{x}{x - 2} \qquad x \neq 2$

You have to exclude $x = 2$ from the domain.

 You can't square root a negative number

$g(x) = \sqrt{x + 3} \qquad x \geqslant -3$

You have to exclude values of x less than −3

Worked example *Aiming higher*

The function f is defined as
$f(x) = \dfrac{1}{x + 7}$

(a) Find the value of f(3) **(1 mark)**

$f(3) = \dfrac{1}{3 + 7} = \dfrac{1}{10}$

(b) State which value(s) of x must be excluded from the domain of f. **(1 mark)**

$x = -7$

For part (b), work out what the possible **output** values could be if the input values are $x \geqslant 2$

Write your answer as f(x) $\geqslant$ ☐

Now try this

Aiming higher

1 The function f is defined as $f(x) = \dfrac{x + 3}{x}$

(a) Find f(1) **(1 mark)**

(b) State which value of x cannot be included in the domain of f. **(1 mark)**

(c) Given that f(b) = 3, find the value of b. **(2 marks)**

2 $f(x) = (x + 1)^2$

(a) Find f(3) **(1 mark)**

(b) The domain of f is all values of x where $x \geqslant 2$ Find the range of f. **(2 marks)**

$g(x) = \sqrt{x - 5}$

(c) Which values of x cannot be included in the domain of g? **(2 marks)**

Composite functions

If you apply two functions one after the other, you can write a **single function** which has the same effect as the two combined functions. This is called a **composite function**.

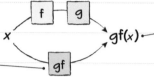

The function gf has the same effect as applying function f **then** applying function g.

The **order** is important. The function being applied **first** goes **closest** to the x.

Order is important in composite functions. You can think of fg(x) as f[g(x)] You work out g(x) **first**, then you use this answer as your **input** for f(x) Note that in this case gf(10) would give you a **different answer**:

$$f(10) = 10^2 = 100$$
$$g(100) = 100 - 3 = 97$$
So gf(10) = 97

Worked example   **Aiming higher**

Worked example   **Aiming higher**

$f(x) = x^2$

$g(x) = x - 3$

Find fg(10)   **(2 marks)**

$g(10) = 10 - 3 = 7$
$f(7) = 7^2 = 49$
So fg(10) = 49

Worked example   **Aiming higher**

$f(x) = 3x + 2$

$g(x) = \dfrac{x}{x + 2}$

Find fg(x)

Give your answer as a single algebraic fraction expressed as simply as possible.   **(3 marks)**

$fg(x) = f[g(x)] = f\left[\dfrac{x}{x + 2}\right]$

$= 3\left[\dfrac{x}{x + 2}\right] + 2$

$= \dfrac{3x}{x + 2} + 2$

$= \dfrac{3x}{x + 2} + \dfrac{2(x + 2)}{x + 2}$

$= \dfrac{3x + 2x + 4}{x + 2} = \dfrac{5x + 4}{x + 2}$

Finding fg(x)

To find an algebraic expression for fg(x) you need to:

1 Write fg(x) as f[g(x)]

2 Substitute the **whole expression** for g(x) for each instance of x in the expression for f(x)
Use square brackets when you substitute.

3 Simplify the new expression as much as possible.
You might have to expand brackets or simplify algebraic fractions.

Now try this   **Aiming higher**

1 The functions f and g are such that
$f(x) = 2x - 2$ and $g(x) = x^2 - 4$
   (a) Find gf(x)
   Give your answer as simply as possible.
   (2 marks)
   (b) Solve gf(x) = 0   **(3 marks)**

In part (a), remember to substitute (x + 1) for **both** instances of x in the expression for g(x)

2 $f(x) = x + 1$
$g(x) = \dfrac{x}{2x - 1}$
   (a) Find gf(x)
   Give your answer as simply as possible.
   (2 marks)
   (b) Find fg(x)
   Give your answer as a single algebraic fraction expressed as simply as possible.
   (3 marks)

Inverse functions

For a function f, the **inverse** of f is the function that **undoes** f. You write the inverse as f^{-1}. If you apply f then f^{-1}, you will end up back where you started.

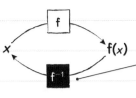

If you apply f, then f^{-1} you have applied the **composite function** $f^{-1}f$. The output of $f^{-1}f$ is the **same** as the input. You can write:

$$f^{-1}f(x) = ff^{-1}(x) = x$$

Finding the inverse

To find the inverse of a function given in the form $f(x) = \dots$ you need to:

 Write the function in the form $y = \dots$

 Rearrange to make x the subject

 Swap each y for an x and rewrite as $f^{-1}(x) = \dots$

For a reminder on changing the subject of a formula, have a look at page 50.

Worked example

Aiming higher

$$g(x) = \frac{3}{x + 4}$$

Express the inverse function g^{-1} in the form $g^{-1}(x) = \dots$ **(3 marks)**

$$y = \frac{3}{x + 4} \qquad (\times (x + 4))$$

$$y(x + 4) = 3 \qquad (\div y)$$

$$x + 4 = \frac{3}{y} \qquad (- 4)$$

$$x = \frac{3}{y} - 4 \qquad \text{(Swap each } y \text{ for an } x)$$

$$g^{-1}(x) = \frac{3}{x} - 4$$

Start by multiplying both sides by $(x + 4)$. You want x on its own on the left-hand side so don't expand the bracket. Once you've rearranged to make x the subject, swap each y for an x and and write your answer as $g^{-1}(x) = \dots$

Worked example

Aiming higher

f is the function $f(x) = 3x + 5$

Express the inverse function f^{-1} in the form $f^{-1}(x) = \dots$ **(2 marks)**

$$y = 3x + 5 \qquad (- 5)$$

$$y - 5 = 3x \qquad (\div 3)$$

$$\frac{y - 5}{3} = x \qquad \text{(Swap each } y \text{ for an } x)$$

$$f^{-1}(x) = \frac{x - 5}{3}$$

You can sometimes use a flow chart to find an inverse.

Here is a flow chart for $f(x)$

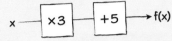

You work **backwards** through the flow chart to find $f^{-1}(x)$

$$f^{-1}(x) \longleftarrow \boxed{\div 3} \boxed{-5} \longleftarrow x$$

You subtract 5, **then** divide by 3. Written using algebra this is

$$f^{-1}(x) = \frac{x - 5}{3}$$

Now try this

Aiming higher

1 f is the function $f(x) = 2x - 1$

(a) Express the inverse function f^{-1} in the form $f^{-1}(x) = \dots$ **(2 marks)**

(b) Without doing any further working, write down the value of $f^{-1}f(7)$ **(1 mark)**

2 The function g is such that $g: x \to \dfrac{x + 6}{x}$

Express the inverse function g^{-1} in the form $g^{-1}: x \to \dots$ **(3 marks)**

Group all the terms involving x on one side, then factorise to get x on its own.

Differentiation

You can **differentiate** a function to find its **derivative** or **gradient function**, $\dfrac{dy}{dx}$

Differentiating x^n

In your International GCSE exam all the functions you have to differentiate will have terms of the form ax^n.

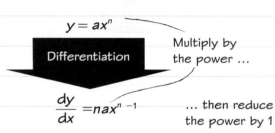

$$y = ax^n$$

Differentiation

Multiply by the power ...

$$\frac{dy}{dx} = nax^{n-1}$$ ... then reduce the power by 1

This rule works for **any** value of n.

Learn this — it's not on the formula sheet.

Golden rules

1 Write every term in the form ax^n before differentiating.

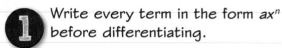
$$y = \frac{1}{x} = x^{-1} \rightarrow \frac{dy}{dx} = -x^{-2}$$

For a reminder about index laws have a look at page 22.

2 Constant terms differentiate to **zero** and x terms differentiate to a **constant**.

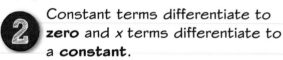
$$y = 9x \rightarrow \frac{dy}{dx} = 9$$

$$y = 25 \rightarrow \frac{dy}{dx} = 0$$

Worked example

Aiming higher

Differentiate with respect to x:

(a) $5x^2$ **(1 mark)**

$$y = 5x^2$$
$$\frac{dy}{dx} = 10x$$

(b) $\dfrac{3}{x}$ **(2 marks)**

$$y = \frac{3}{x} = 3x^{-1}$$
$$\frac{dy}{dx} = -3x^{-2}$$

'With respect to x' just means that x is the variable. If there were any other letters, you would treat them as **constants**.

(a) You multiply by the power, then reduce the power by 1.
$$\frac{dy}{dx} = 2 \times 5x^{2-1} = 10x$$

(b) You need every term to be in the form ax^n before you differentiate. Start by rewriting $\dfrac{3}{x}$ as $3x^{-1}$. Then multiply by the power and reduce the power by 1:
$$\frac{dy}{dx} = -1 \times 3x^{-1-1} = -3x^{-2} \text{ or } \frac{-3}{x^2}$$

Differentiate **term by term**. Be careful with the signs. Remember that the x term gives a **constant** so differentiating $-6x$ gives -6. Any constant terms go to **zero**, so the $+10$ term disappears when you differentiate.

Worked example

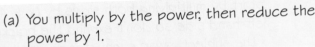
Aiming higher

For the curve with equation $y = 2x^3 - 6x + 10$, find $\dfrac{dy}{dx}$ **(2 marks)**

$$y = 2x^3 - 6x + 10$$
$$\frac{dy}{dx} = 6x^2 - 6$$

Now try this

Aiming higher

1 Differentiate with respect to x
   (a) $4x^2 + 6x - 1$ **(2 marks)**
   (b) $\dfrac{2}{x}$ **(2 marks)**

2 (a) Write $\dfrac{1+x}{x^2}$ in the form $x^a + x^b$ where a and b are integers. **(2 marks)**
   (b) Differentiate $\dfrac{1+x}{x^2}$ with respect to x. **(2 marks)**

Gradients and calculus

You can use the **derivative** or **gradient function** to find the **rate of change** of a function, or the gradient of a curve.

This curve has equation $y = x^3 + 5x^2$. Its gradient function has equation $\dfrac{dy}{dx} = 3x^2 + 10x$. You can find the **gradient** at any point on the graph by substituting the x-coordinate at that point into the gradient function.

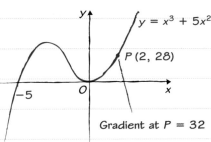

$y = x^3 + 5x^2$

$P(2, 28)$

-5

Gradient at $P = 32$

At the point P, $x = 2$, so $\dfrac{dy}{dx} = 3(2)^2 + 10(2) = 12 + 20 = 32$ The gradient at P is 32.

Worked example
Aiming higher

For the curve with equation $y = x^3 - 5x + 7$

(a) find $\dfrac{dy}{dx}$ **(2 marks)**

$y = x^3 - 5x + 7$

$\dfrac{dy}{dx} = 3x^2 - 5$

(b) find the gradient of the curve at the point where $x = 5$ **(2 marks)**

When $x = 5$, $\dfrac{dy}{dx} = 3(5)^2 - 5$

$= 75 - 5 = 70$

(a) Differentiate **term by term** to find $\dfrac{dy}{dx}$
For a reminder about differentiating have a look at page 58.

(b) The value of $\dfrac{dy}{dx}$ when $x = 5$ tells you the **gradient** of the **original curve** when $x = 5$. Substitute $x = 5$ into the expression for $\dfrac{dy}{dx}$

Problem solved!

In this question you are given the **gradient** and you have to find the **coordinates** of the points with that gradient. Follow these steps:

1. Differentiate to find the gradient function.
2. Set the gradient function equal to 48.
3. **Solve** the equation to find the values of x with that gradient. Be careful – there may be more than one solution.
4. Substitute those values of x into the **original** equation to find the corresponding values of y.

You will need to use problem-solving skills throughout your exam – **be prepared!**

Worked example
Aiming higher

Find the coordinates of the points on the curve $y = x^3$ where the gradient is 48.   **(3 marks)**

$y = x^3$

$\dfrac{dy}{dx} = 3x^2$

When $\dfrac{dy}{dx} = 48$, $3x^2 = 48$ $(\div 3)$

$x^2 = 16$ $(\sqrt{\ })$

$x = 4$ or $x = -4$

If $x = 4$, $y = 4^3 = 64$

If $x = -4$, $y = (-4)^3 = -64$

The two points are $(4, 64)$ and $(-4, -64)$

Now try this

1 For the curve with equation $y = 3x^3 + 2x^2 + 5$

 Aiming higher

  (a) find $\dfrac{dy}{dx}$ **(2 marks)**

  (b) find the gradient of the curve at the point where $x = 2$ **(2 marks)**

2 Find the coordinates of the point on the curve $y = 5x^2 + 3x + 2$ where the gradient is 11   **(3 marks)**

3 For the curve with equation $2x^3 - 6x + 1$

  (a) find $\dfrac{dy}{dx}$ **(2 marks)**

  (b) find the coordinates of the two points on the curve where the gradient of the curve is 0   **(4 marks)**

Set $\dfrac{dy}{dx} = 0$ to find **two** values of x.

Turning points and calculus

You can use **differentiation** to find the **turning points** and **stationary points** of a **graph** or **function** in your International GCSE exam. You need to be confident with differentiation so revise that on pages 58 and 59 first.

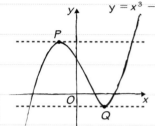

$y = x^3 - 5x + 3$

This graph has turning points at P and Q. The gradient is O at both points.

Golden rule

The turning points and stationary points of a graph or function are the points where the **derivative**, or **gradient function**, $\dfrac{dy}{dx}$, is equal to **zero**.

Worked example   *Aiming higher*

Find the coordinates of the turning point on the curve with equation $y = 3x^2 + 12x + 5$ **(4 marks)**

$$\frac{dy}{dx} = 6x + 12$$

When $\dfrac{dy}{dx} = 0$, $6x + 12 = 0$

$$6x = -12$$

$$x = -2$$

So $y = 3 \times (-2)^2 + 12 \times (-2) + 5 = -7$

Turning point is $(-2, -7)$

To find the coordinates of the turning point using differentiation:

1. Differentiate to find $\dfrac{dy}{dx}$
2. Set $\dfrac{dy}{dx} = 0$
3. Solve the equation to find the value or values of x.
4. Find the corresponding value of y for each value of x.

Maximum or minimum?

For a **quadratic** function, the turning point is either a maximum or a minimum. You can work out which using the **coefficient** of x^2.

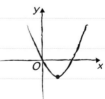   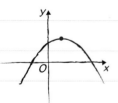

If the coefficient of x^2 is **positive** the turning point is a **minimum**.

If the coefficient of x^2 is **negative** the turning point is a **maximum**.

Worked example   *Aiming higher*

For $A = 6x^2 - 90x + 10$

(a) find $\dfrac{dA}{dx}$ **(2 marks)**

$$\frac{dA}{dx} = 12x - 90$$

(b) find the value of x for which A is a minimum   **(2 marks)**

When $\dfrac{dA}{dx} = 0$, $12x - 90 = 0$ $(+ 90)$

$$12x = 90 \quad (\div 12)$$

$$x = 7.5$$

(c) explain how you know that A is a minimum for this value of x.   **(1 mark)**

Because the coefficient of x^2 is positive.

Now try this

 Aiming higher

1 (a) Find the coordinates of the turning point on the curve with equation $y = x^2 - 8x + 3$ **(4 marks)**

(b) State with a reason whether this turning point is a minimum or a maximum.   **(1 mark)**

2 For the curve with equation $y = x^3 - 5x^2 + 8x + 1$

(a) find $\dfrac{dy}{dx}$ **(2 marks)**

(b) find the x-coordinates of the two stationary points on the curve.   **(4 marks)**

Kinematics

Kinematics deals with the **motion** of objects. You can use **differentiation** to solve problems involving the **displacement**, **velocity** and **acceleration** of an object.

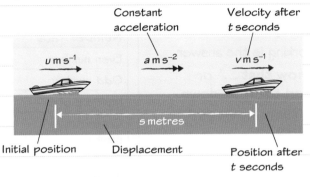

Constant acceleration

Velocity after t seconds

$u\,m\,s^{-1}$ $a\,m\,s^{-2}$ $v\,m\,s^{-1}$

s metres

Initial position   Displacement   Position after t seconds

Worked example

Aiming higher

A particle moves in a straight line through a fixed point O. The displacement of the particle from O at time t seconds is s metres, where

$$s = t^2 - 6t + 10$$

(a) Find $\dfrac{ds}{dt}$ **(2 marks)**

$$\frac{ds}{dt} = 2t - 6$$

(b) Find the velocity of particle when $t = 5$ **(2 marks)**

$$v = \frac{ds}{dt} = 2t - 6$$

When $t = 5$, $v = 2(5) - 6$

$$= 10 - 6$$
$$= 4 \text{ m/s}$$

(c) Find the acceleration of the particle. **(2 marks)**

$$v = 2t - 6$$
$$a = \frac{dv}{dt} = 2 \text{ m/s}^2$$

Differentiating with respect to time

The derivative tells you the **rate of change** of one variable with respect to another.

Displacement, s

Differentiation

Velocity, v = rate of change of displacement with time = $\dfrac{ds}{dt}$

Differentiation

Acceleration, a = rate of change of velocity with time = $\dfrac{dv}{dt}$

Differentiate the expression for s once to find an expression for v. The units for velocity are metres per second (m/s). Differentiate again to find an expression for a. The units for acceleration are metres per second per second (m/s²).

Now try this

Aiming higher

A stone is projected vertically upwards from the ground. After t seconds its height above the ground, s metres, is given by

$$s = 15t - 4.9t^2 \quad \text{for } 0 \leqslant t \leqslant 4$$

(a) Find $\dfrac{ds}{dt}$ **(2 marks)**

(b) Find the velocity of the stone when $t = 0.5$ **(1 mark)**

(c) Find the maximum height of the stone above the ground, correct to 1 decimal place. **(3 marks)**

Set $\dfrac{ds}{dt} = 0$ and solve to find the value of t which gives a **maximum** value of s. Then substitute this value of t back into the equation for s to find the maximum height.

Algebraic proof

You can use algebra to **prove** facts about numbers.

Using algebra helps you to prove that something is true for **every** number.

In a proof question the working **is** the answer.

If the question says '**Show that** ...' or '**Prove that** ...', you need to write down every stage of your working. If you don't, you won't get all the marks.

Golden rule

If you need to prove something about numbers then you always use algebra.

Algebraic proof toolkit

Use n to represent any whole number.

Number fact	Written using algebra
Even number	$2n$
Odd number	$2n + 1$ or $2n - 1$
Multiple of 3	$3n$
Consecutive numbers	$n, n + 1, n + 2, \ldots$
Consecutive even numbers	$2n, 2n + 2, 2n + 4, \ldots$
Consecutive odd numbers	$2n + 1, 2n + 3, 2n + 5, \ldots$
Consecutive square numbers	$n^2, (n + 1)^2, (n + 2)^2, \ldots$

Problem solved!

To **prove** the statement you need to show that it is true for **any** three consecutive integers. You can do this using algebra.
1. Write the first integer as n and the next two integers as $n + 1$ and $n + 2$.
2. Write an expression for the sum of your three integers.
3. Simplify and factorise the expression.
4. Explain why the final expression is divisible by 3.

> You will need to use problem-solving skills throughout your exam – **be prepared!**

Worked example

Aiming higher

Prove that the sum of three consecutive integers is always divisible by 3. **(3 marks)**

n, $n + 1$ and $n + 2$ represent any three consecutive integers.

$n + (n + 1) + (n + 2) = 3n + 3$
$= 3(n + 1)$

$n + 1$ is an integer, so $3(n + 1)$ is divisible by 3.

Identities

The symbol $\equiv$ represents an **identity**. This is something which is always true. If you have to show that an identity is true in your exam:
- start with the left-hand side
- manipulate it until it matches the right-hand side.

Worked example

Show that $(a + b)(a - b) \equiv a^2 - b^2$ **(2 marks)**

$(a + b)(a - b) = a^2 - ab + ab - b^2$
$= a^2 - b^2$

> An identity is not an equation – don't try to solve it using the balance method.

Now try this

Worked solution video

 1 The nth term for the sequence of triangular numbers is
$\frac{1}{2}n(n + 1)$
Prove that the sum of two consecutive triangular numbers is a square number. **(3 marks)**

> Write the two integers as n and $(n + 4)$.

2 Two integers have a difference of 4. The difference between the squares of the two integers is 8 times the mean of the integers.
For example, $14 - 10 = 4$
$14^2 - 10^2 = 196 - 100 = 96$

The mean of 10 and 14 is $\frac{10 + 14}{2} = 12$
and $8 \times 12 = 96$

Prove this result algebraically. **(4 marks)**

Problem-solving practice 1

In your International GCSE Maths exams you will need to demonstrate **problem-solving** and **reasoning skills**. If you come across a tricky or unfamiliar question in your exam, you can try some of these strategies:

- ✓ Sketch a diagram to see what is going on.
- ✓ Try the problem with smaller or easier numbers.
- ✓ Plan your strategy before you start.
- ✓ Write down any formulae you might be able to use, or check the formulae sheet for a clue.
- ✓ Use x or n to represent an unknown value.

---

1 Here are the first four terms of an arithmetic sequence.

9, 15, 21, 27, ...

Explain why the number 65 cannot be a term of this sequence. **(3 marks)**

Worked solution video

Arithmetic sequences page 25

Work out the nth term of the sequence. You could set it equal to 65 and solve an equation to show that n is not an integer, or you could work out consecutive terms of the sequence on either side of 65.

TOP TIP

You can use any successful strategy to answer a question, as long as you show your method.

---

2 A train travels a distance of y km in 3 minutes. Show that the average speed of the train is $20y$ km/h. **(3 marks)**

Formulae page 24
Speed page 65

Watch out! The time is given in minutes, but you need to find an expression for the speed in km/h. Divide by 60 then simplify your fraction. Because this is a 'Show that...' question you need to write down your conversion **clearly** and give the **units**.

TOP TIP

Try changing y into a number and calculating the average speed. It will give you an idea of how the formula works.

---

3 A straight line has gradient 8 and passes through the point (5, 10).

(a) Find the equation of the line. **(2 marks)**

(b) Determine whether the point (10, 50) lies on the line. **(1 mark)**

Straight-line graphs 2 page 27

You could tackle this question using algebra, or by drawing a diagram to scale. Whichever method you use, make sure you write a conclusion.

TOP TIP

If you have to comment on **reliability** (i.e. how reliable something is) then think about the method you used. If you used a diagram your answer might be less reliable than if you used algebra.

Problem-solving practice 2

4 Factorise completely
$(6a + b)^2 - (a - 3b)^2$

(2 marks)

Factorising page 21

This is in the form $x^2 - y^2$ with $x = 6a + b$ and $y = a - 3b$

Use this rule:

$(x + y)(x - y) = x^2 - y^2$

TOP TIP

If you have to factorise an expression, always keep an eye out for the difference of two squares.

---

5 (a) Show that $\dfrac{x^2 - 2x}{3x^2 - 5x - 2}$ can

be written as $\dfrac{x}{kx + 1}$ and state

the value of k.   **(3 marks)**

(b) $f(x) = \dfrac{x}{3x + 1}$

Find the inverse function f^{-1} in the form $f^{-1}(x) = ...$, showing your working clearly.

(3 marks)

Algebraic fractions page 52 · Aiming higher
Inverse functions page 57

In part (a), you need to **factorise** the top and bottom of the fraction before cancelling one of the factors.

TOP TIP

Finding an inverse of a function is the same skill as **rearranging formulae**. Check you are confident with that on page 51.

---

6 At an open-air sale, two plots are marked out using tape. ABFE is a rectangle of width x m and length y m. The line CD divides it into two equal parts. The total length of tape used is 60 m.

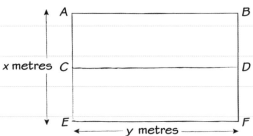

(a) (i) Show that $y = 20 - \dfrac{2}{3}x$

The area of the rectangle ABFE is A m^2

(ii) Show that $A = 20x - \dfrac{2}{3}x^2$ **(3 marks)**

(b) Find $\dfrac{dA}{dx}$ **(2 marks)**

(c) Find the maximum value of A.   **(3 marks)**

Differentiation page 58 · Aiming higher
Turning points page 60

(a) You are given the length of tape used. The diagram shows you that you need 3 lots of y m and 2 lots of x m. Use this information to write down the relationship between x and y using algebra, then rearrange your equation until it matches the one given.

(b) When you differentiate, the variables don't have to be y and x. Here they are A and x.

(c) The maximum value for A occurs at the **turning point**. Set $\dfrac{dA}{dx}$ equal to 0 and solve an equation to find A.

TOP TIP

If you get stuck on an early part of a long question, see if you have enough information to tackle a later part. In this question you could do part (b) even if you haven't completed part (a).

Speed

This is the formula triangle for speed.

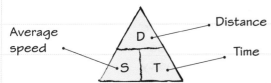

Average speed, Distance, Time

Average speed = $\dfrac{\text{total distance travelled}}{\text{total time taken}}$

Time = $\dfrac{\text{distance}}{\text{average speed}}$ **LEARN IT!**

Distance = average speed × time

Using a formula triangle

Cover up the quantity you want to find with your finger.

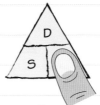

The position of the other two quantities tells you the formula.

$T = \dfrac{D}{S}$ $S = \dfrac{D}{T}$ $D = S \times T$

Units

The most common units of speed are:

- metres per second: m/s
- kilometres per hour: km/h
- miles per hour: mph.

To convert between measures of speed you need to convert one unit first then the other. Write the new units at each step of your working. To convert 72 km/h into m/s:

$72\,\text{km/h} \rightarrow 72 \times 1000 = 72\,000\,\text{m/h}$

$72\,000\,\text{m/h} \rightarrow 72\,000 \div 3600 = 20\,\text{m/s}$

1 hour = 60 × 60 = 3600 seconds

Minutes and hours

For questions on speed, you need to be able to convert between minutes and hours.

Remember there are 60 minutes in 1 hour.

To convert from minutes to hours you divide by 60.

24 minutes = 0.4 hours $\dfrac{24}{60} = \dfrac{2}{5} = 0.4$

To convert from hours to minutes you multiply by 60.

0.2 hours = 12 minutes $3.2 \times 60 = 192$

3.2 hours = 3 hours 12 minutes

Worked example

The speed of light in a vacuum is approximately 1.08×10^9 km/h.

Light from the Sun takes approximately 8 minutes and 15 seconds to travel to Earth.

Estimate the distance from the Earth to the Sun.   **(3 marks)**

8 mins 15 secs = 8.25 mins = $\dfrac{8.25}{60}$
  = 0.1375 hours

$D = S \times T$
  $= 1.08 \times 10^9 \times 0.1375$
  $= 1.485 \times 10^8$ km

Be careful with the units. You need to convert 12 minutes and 15 minutes into hours before doing your calculations.

Speed checklist

Draw formula triangle.
Make sure units match.   ✓
Give units with answer.   ✓

If you're answering questions involving speed, distance and time you must always make sure that the units match. Speed is given in km/h here, so convert the time into hours before calculating.

Now try this

Rosa lives in Durham and works in Newcastle. She takes the train to work every day.

Last Tuesday her train journey to work took 12 minutes, at an average speed of 108 km/h.

Her journey home from work took 15 minutes.

Calculate Rosa's average speed on her journey home.   **(3 marks)**

Density

The density of a material is its mass per unit volume.

This is the formula triangle for density.

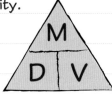

$$\text{Density} = \frac{\text{mass}}{\text{volume}}$$

$$\text{Volume} = \frac{\text{mass}}{\text{density}}$$

$$\text{Mass} = \text{density} \times \text{volume}$$

Worked example

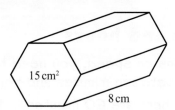

15 cm²

8 cm

The diagram shows a solid hexagonal prism.
The area of the cross-section of the prism is 15 cm².
The length of the prism is 8 cm.
The prism is made from wood with a density of 0.8 grams per cm³.
Work out the mass of the prism.   **(4 marks)**

Volume of prism
$$= \text{area of cross-section} \times \text{length}$$
$$= 15 \times 8$$
$$= 120 \, \text{cm}^3$$
$$M = D \times V$$
$$= 0.8 \times 120$$
$$= 96$$
The mass of the prism is 96 g.

Revise volumes of prisms on page 73.

Units
The most common units of density are:
- grams per cubic centimetre: g/cm^3
- kilograms per cubic metre: kg/m^3

Mass = density × volume

You are given the density so you need to work out the volume of the prism.

You need to learn the formula for the volume of a prism.

The density is in grams per cm³ and the volume is in cm³ so the mass will be in grams.

Worked example

An iron bar has a volume of 1.2 m³ and a mass of 9444 kg. Calculate the density of iron.   **(2 marks)**

$$D = \frac{M}{V} = \frac{9444}{1.2} = 7870 \, \text{kg/m}^3$$

Volume is in m³ and mass is in kg so density will be in kg/m^3.

Worked solution video

Now try this

The density of copper is 8.92 g/cm³.
The density of silver is 10.49 g/cm³.
20 cm³ of copper and 5 cm³ of silver are mixed together to make a new kind of metal.
Work out the density of the new metal.

          (4 marks)

Plan your answer:
1. Work out the mass of each metal.
2. Add the masses together.
3. You know the total mass and the total volume of the new metal, so calculate the density.

Other compound measures

Compound measures are made up of two or more other measurements. **Speed** is a compound measure because it is calculated using distance **and** time. **Density** is a compound measure because it is calculated using mass **and** volume. You need to be able to work with other compound measures as well.

Pressure

Pressure is a measure of the force applied over a given area. The most common units of pressure are newtons per square centimetre (N/cm^2) and newtons per square metre (N/m^2). You can use the formula triangle on the right to calculate with pressure.

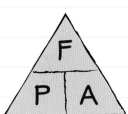

$$Pressure = \frac{Force}{Area}$$

$$Area = \frac{Force}{Pressure}$$

$$Force = Pressure \times Area$$

LEARN IT!

Worked example

At a depth of 15 m, water has a pressure of $14.7\,N/cm^2$. Calculate the force applied to a diving mask with a surface area of $360\,cm^2$.   **(2 marks)**

Force = Pressure × Area
    = 14.7 × 360
    = 5292 N

Rates

If the 'bottom' unit in a compound measure is **time**, then it is a **rate**. Here are some examples.

$$Speed = \frac{Distance}{Time}$$

$$Rate\ of\ flow = \frac{Volume}{Time}$$

$$Rate\ of\ climb = \frac{Height}{Time}$$

$$Rate\ of\ pay = \frac{Salary}{Time}$$

Problem solved!

You need to use lots of different maths skills to solve this problem. You need to know how to calculate the volume of a cuboid and how to convert from cm^3 to litres (for a reminder of both look at page 79). It's a good idea to plan your strategy before you start:

1. Work out the volume of the fish tank.
2. Convert from cm^3 into litres.
3. Divide by 2 to get half the capacity.
4. Divide by the rate of flow to get the time taken in minutes.

> You will need to use problem-solving skills throughout your exam – **be prepared!**

Worked example

This fish tank can be modelled as a cuboid.

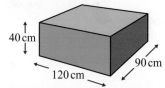

40 cm   90 cm   120 cm

Antonio uses a hose to fill the fish tank with water at a rate of 12 litres per minute. How long will it take for the fish tank to be half full?   **(5 marks)**

Volume = 120 × 90 × 40
    = 432 000 cm³
    = 432 litres
432 ÷ 2 = 216
216 ÷ 12 = 18
It will take 18 minutes for the tank to be half full.

Now try this

1 The average fuel consumption of a car is measured in kilometres per litre (km/l). A car travels 249 km and uses 15 litres of petrol. What is its average fuel consumption?   **(2 marks)**

Look at the units to work out what calculation to do.

2 A large tank holds 1680 litres of water. The tank can be filled from a hot tap or from a cold tap. The cold tap on its own takes 4 minutes to fill the tank. The hot tap on its own takes 6 minutes to fill the tank.

Preti turns both taps on at the same time. How long does the tank take to fill?   **(3 marks)**

Angle properties

You need to remember all of these angle properties and their correct names.

Corresponding angles are equal.

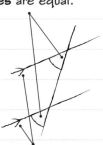

Parallel lines are marked with arrows.

Alternate angles are equal.

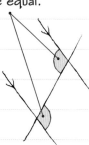

Vertically opposite angles are equal.

Co-interior or allied angles add up to 180°.

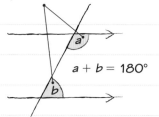

$a + b = 180°$

These are useful angle facts for triangles and parallelograms:

Interior angle

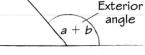

The exterior angle of a triangle is equal to the sum of the interior angles at the other two vertices.

Exterior angle

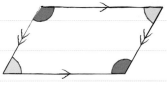

The opposite angles of a parallelogram are equal.

Golden rule

When answering angle problems, you need to give a reason for each step of your working.

Angle sums

You need to remember these two angle facts:

1 The angles in a triangle add up to 180°.

2 The angles in a quadrilateral add up to 360°.

Worked example

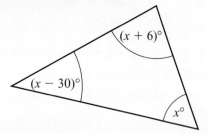

Work out the value of x.   **(3 marks)**

Angles in a triangle add up to 180° so

$(x - 30) + (x + 6) + x = 180$

$3x - 24 = 180$ $(+ 24)$

$3x = 204$ $(÷ 3)$

$x = 68$

Use the fact that the angles in a triangle add up to 180° to write an equation, then solve your equation to find x. For a reminder about solving linear equations have a look at page 22.

Now try this

AB is parallel to CD.

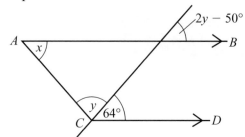

Work out the value of x.   **(5 marks)**

Solving angle problems

You might need to use angle properties to solve problems in your exam. Remember to give reasons for every step of your working.

Reasons

Use these reasons in angle problems:
- Angles on a straight line add up to 180°.
- Angles around a point add up to 360°.
- Opposite angles are equal.
- Corresponding angles are equal.
- Co-interior angles add up to 180°.
- Alternate angles are equal.
- Angles in a triangle add up to 180°.
- Angles in a quadrilateral add up to 360°.
- Base angles of an isosceles triangle are equal.

Use the properties on the diagram:
AB is parallel to CD
AC is parallel to BD
BE is equal to DE

Worked example

Work out the size of the angle marked x.
Give reasons for each step of your working.

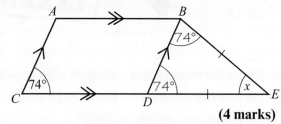

(4 marks)

$\angle BDE = 74°$ (corresponding angles are equal)

$\angle DBE = 74°$ (base angles in an isosceles triangle are equal)

$x + 74° + 74° = 180°$ (angles in a triangle add up to 180°)

$x = 180° - 148°$

$x = 32°$

Worked example

ABCD is a quadrilateral.
Angle DAB is a right angle.
Angles ABC and BCD are in the ratio $1:2$
Angle CDA is 70° more than angle ABC.
Work out the size of angles ABC, BCD and CDA. **(4 marks)**

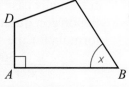

$A = 90°$

$C = 2x$

$D = x + 70°$

$A + B + C + D = 360°$ (angles in a quadrilateral add up to 360°)

So $90° + x + 2x + (x + 70°) = 360°$

$4x = 200°$

$x = 50°$

So $B = 50°$, $C = 2 \times 50° = 100°$ and $D = 50° + 70° = 120°$

Write angle B as x then express angles C and D in terms of x.
Angle problems could involve ratio or proportion, or you might need to write your own equation.

Now try this

In the diagram ABC and BDC are isosceles triangles.
Express the size of angle ABD in terms of x, giving your answer as simply as possible.
Give a reason for each step of your working.

(4 marks)

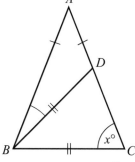

Angles in polygons

Polygon questions are all about interior and exterior angles.

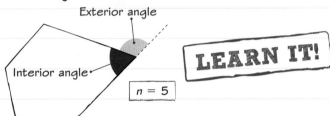

Exterior angle

Interior angle

$n = 5$

LEARN IT!

This diagram shows part of a **regular** polygon with 30 sides.

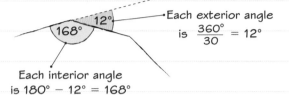

168°   12°

Each exterior angle is $\frac{360°}{30} = 12°$

Each interior angle is $180° - 12° = 168°$

Use these formulae for a polygon with n sides.

Sum of interior angles $= 180° \times (n - 2)$

Sum of exterior angles $= 360°$

Don't try to draw a 30-sided polygon!

If there's no diagram given in a polygon question, you probably don't need to draw one.

Regular polygons

In a regular polygon all the sides are equal and all the angles are equal.

If a regular polygon has n sides then each exterior angle is $\frac{360°}{n}$

LEARN IT!

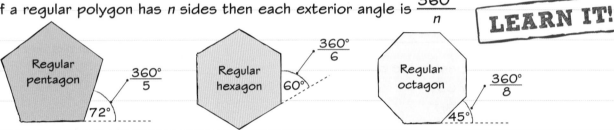

Regular pentagon $\frac{360°}{5}$ 72°

Regular hexagon   60° $\frac{360°}{6}$

Regular octagon $\frac{360°}{8}$ 45°

You can use the fact that angles on a straight line add up to 180° to work out the size of one of the interior angles.

Worked example

The diagram shows part of a regular polygon. The interior angle and the exterior angle at a vertex are marked.

The size of the interior angle is 7 times the size of the exterior angle.

Work out the number of sides of the polygon.   **(3 marks)**

$180° \div 8 = 22.5°$

$\frac{360°}{22.5°} = 16$

The polygon has 16 sides.

Problem solved!

It's usually easier to work with **exterior** angles in polygon questions. You can rearrange the formula for the size of an exterior angle to get:

$$n = \frac{360°}{\text{exterior angle}}$$

You can use ratios to answer this question. The ratio of the interior angle to the exterior angle is $7:1$. These angles add up to 180° so divide 180° in the ratio $7:1$ to find the exterior angle.

You will need to use problem-solving skills throughout your exam – **be prepared!**

Now try this

The diagram shows part of a regular polygon with n sides.
(a) Work out the value of n.   **(2 marks)**
(b) What is the sum of the interior angles of the polygon?   **(2 marks)**

10°

Worked solution video

Perimeter and area

Triangle

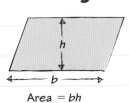

Area $= \frac{1}{2}bh$

Parallelogram

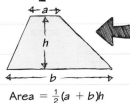

Area $= bh$

Trapezium

Area $= \frac{1}{2}(a + b)h$

> The formula for the area of a trapezium is given on the formulae sheet, but you need to **learn** the formulae for a triangle and a parallelogram.

You can calculate areas and perimeters of more complex shapes by splitting them into parts.

You might need to draw some extra lines on your diagram and add or subtract areas.

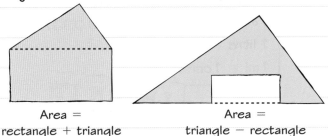

Area = rectangle + triangle

Area = triangle − rectangle

Area basics

- ✓ Lengths are all in the same units.
- ✓ Give units with the answer.
- ✓ Lengths in cm means area units are cm^2.
- ✓ Lengths in m means area units are m^2.

Worked example

This diagram shows a trapezium with base 6 cm and height x cm.

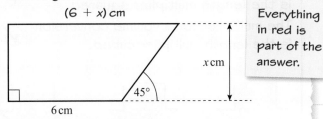

(6 + x) cm

x cm

45°

6 cm

> Everything in red is part of the answer.

The area of the trapezium is A cm^2.

(a) Write a formula for A in terms of x. Give your answer in its simplest form. **(3 marks)**

Area $= \frac{1}{2}(a + b)h$

$A = \frac{1}{2}(6 + 6 + x)x = 6x + \frac{1}{2}x^2$

(b) The height of the trapezium is 5 cm. Find the area of the trapezium. **(2 marks)**

$A = 6 \times 5 + \frac{1}{2} \times 5^2 = 42.5 \, cm^2$

Problem solved!

You will probably have to solve unfamiliar problems involving area and perimeter. Here are some top tips:

- If you need to use a formula, write it out before you substitute any values.
- Write any lengths or angles you work out neatly on the diagram given.
- Draw your own sketch if that helps.
- Don't be afraid to use algebra – you can substitute variables into a formula as well as numbers.
- If it's tricky, try using some numbers instead of variables first to see what is going on.

In this question you have to spot that the 45° angle makes it easy to work out the length of the top of the trapezium.

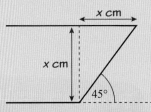

x cm

x cm

45°

> You will need to use problem-solving skills throughout your exam – **be prepared!**

Now try this

The diagram shows a triangle.
In the diagram, all the measurements are in centimetres.
The area of the triangle is A cm^2.
Work out the value of A. **(4 marks)**

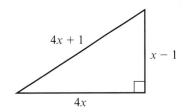

$4x + 1$

$x - 1$

$4x$

Worked solution video

Units of area and volume

Converting units of area or volume is trickier than converting units of length. You need to remember your area and volume conversions for your exam.

These two squares have the same area.

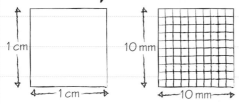

1 cm   10 mm

1 cm   10 mm

So $1\,cm^2 = 100\,mm^2$.

These two cubes have the same volume.

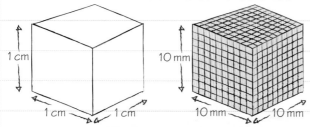

1 cm   1 cm   1 cm   10 mm   10 mm   10 mm

So $1\,cm^3 = 1000\,mm^3$.

Area conversions

$1\,cm^2 = 10^2\,mm^2 = 100\,mm^2$

$1\,m^2 = 100^2\,cm^2 = 10\,000\,cm^2$

$1\,km^2 = 1000^2\,m^2 = 1\,000\,000\,m^2$

Volume conversions

$1\,cm^3 = 10^3\,mm^3 = 1000\,mm^3$

$1\,m^3 = 100^3\,cm^3 = 1\,000\,000\,cm^3$

1 litre $= 1000\,cm^3$

$1\,ml = 1\,cm^3$

LEARN IT!

Worked example

Lead has a density of $11\,350\,kg/m^3$.
An antique lead model has a volume of $400\,cm^3$.
Calculate the mass of the model in kg.

(3 marks)

$400 \div 100^3 = 0.0004$

Volume $= 0.0004\,m^3$

Mass $=$ Density $\times$ Volume
$= 11\,350 \times 0.0004$
$= 4.54\,kg$

Unit conversion checklist

The multiplier for an area conversion is the length multiplier squared. ✓

The multiplier for a volume conversion is the length multiplier cubed. ✓

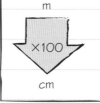 

m m^2 m^3

$\times 100$ $\times 100^2$ $\times 100^3$

cm cm^2 cm^3

Problem solved!

You need to be really careful with the **units** when you are solving any problem involving measures. The units of density are given in kg/m^3, so you need to convert $400\,cm^3$ into m^3 before you calculate. You are converting to a larger unit, so **divide** by 100^3.

For a reminder about density look at page 66.

> You will need to use problem-solving skills throughout your exam – **be prepared!** 💡

Now try this

Worked solution video

1 Convert
(a) $2.3\,m^2$ into cm^2 **(1 mark)**
(b) $400\,mm^3$ into cm^3 **(1 mark)**

2 Convert $0.35\,m^3$ to mm^3, giving your answer in standard form.   **(2 marks)**

> For a reminder about writing answers in standard form have a look at page 16.

3 Dhruv applies a force of $600\,N$ to the floor. The total area of his feet is $160\,cm^2$. What is the pressure, in N/m^2, between him and the floor if he stands on both of his feet? **(3 marks)**

Prisms

Volume

A prism is a 3D solid with a **constant** cross-section. Use this formula to calculate the volume of a prism. It is given on the formulae sheet.

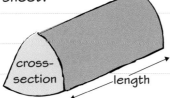

cross-section
length

Volume = area of cross-section × length

Worked example

The diagram shows a prism. The cross-section is a trapezium. Work out the volume of the prism.

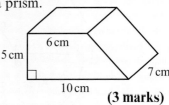

6 cm
5 cm
7 cm
10 cm

(3 marks)

Area of cross-section (trapezium)
$$= \frac{1}{2} \times (6 + 10) \times 5 = 40\,\text{cm}^2$$
Volume of prism = 40 × 7 = 280 cm³

Surface area

To work out the surface area of a 3D shape, you need to add together the areas of all the faces.

It's a good idea to sketch each face with its dimensions.

Remember to include the faces that you can't see.

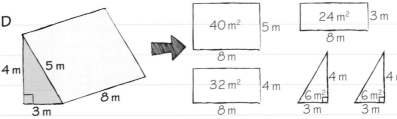

4 m   5 m
3 m   8 m

40 m²   5 m   24 m²   3 m
        8 m   8 m
32 m²   4 m   4 m   4 m
        8 m   6 m²   6 m²
               3 m   3 m

Surface area = 40 + 32 + 24 + 6 + 6 = 108 m²

Worked example

The diagram shows a triangular prism and a cube. They both have the **same** volume. Work out the length of x.   **(4 marks)**

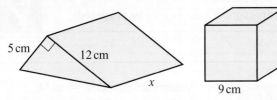

5 cm
12 cm
x
9 cm

Volume of cube = 9³ = 729 cm³

Volume of prism = Area of cross-section
                  × length
$$= \frac{1}{2} \times 5 \times 12 \times x$$
$$= 30x$$
$$30x = 729$$
$$x = 24.3\,\text{cm}$$

Problem solved!

Calculate the volume of the cube, and write an expression for the volume of the prism. Set these equal to each other and solve the equation to find x.

You will need to use problem-solving skills throughout your exam – **be prepared!**

Use this formula to work out the areas of the triangular faces:
Area of a triangle = $\frac{1}{2}$ × base × vertical height

Now try this

The cross-section of this prism is a right-angled triangle.
(a) Work out the volume of the prism.   **(3 marks)**
(b) Work out the total surface area of the prism.   **(3 marks)**

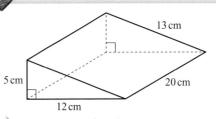

13 cm
5 cm   20 cm
12 cm

Circles and cylinders

You need to learn the formulae for the area and circumference of a circle.

Circle

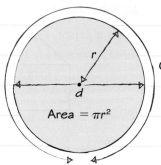

Circumference $= 2\pi r$
$= \pi d$

Area $= \pi r^2$

Cylinder

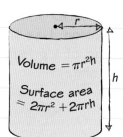

Volume $= \pi r^2 h$

Surface area $= 2\pi r^2 + 2\pi rh$

LEARN IT!

Worked example

The diagram shows a game counter in the shape of a semicircle.

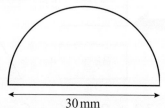

30 mm

Work out the area of the counter. Give your answer correct to 2 significant figures. **(3 marks)**

Radius $= 30 \div 2 = 15$ mm
Area of circle $= \pi r^2$
$= \pi \times 15^2$
$= 706.8583...$ mm^2
Area of counter $= 706.8583... \div 2$
$= 350$ mm^2 (2 s.f.)

Calculator skills

Make sure you know how to enter π on your calculator. On some calculators you have to press these keys:

$\pi \quad e$

SHIFT ➡ $\times 10^x$

Problem solved!

The formula for the area of a circle uses the **radius**. If the length shown on the diagram is the **diameter**, you need to divide it by 2 before you substitute into the formula. Don't round any values until the end of your working.

You will need to use problem-solving skills throughout your exam – **be prepared!**

In terms of π

Unless a question asks you for a specific degree of accuracy, you can give your answers as a whole number or fraction multiplied by π. An answer given in terms of π is an **exact answer** rather than a **rounded answer**.

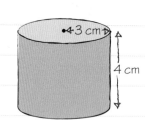

3 cm

4 cm

Volume of cylinder $= \pi r^2 h$
$= \pi \times 3^2 \times 4$

Exact answer
Volume $= 36\pi$ cm^3

Rounded answer
Volume $= 113$ cm^3 (to 3 s.f.)

Now try this

This shape is made from a rectangle and a semicircle.
Work out the area of the shape.
Give your answer correct to 3 significant figures. **(4 marks)**

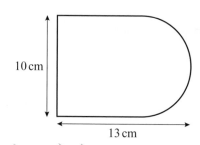

10 cm

13 cm

Sectors of circles

Each pair of radii divides a circle into two sectors, a **major sector** and a **minor sector**.

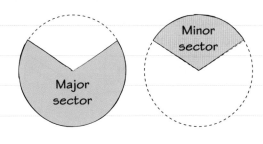

You can find the area of a sector by working out what fraction it is of the whole circle.

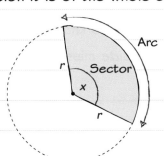

For a sector with angle x of a circle with radius r:

Sector $= \dfrac{x}{360°}$ of the whole circle so

Area of sector $= \dfrac{x}{360°} \times \pi r^2$

Arc length $= \dfrac{x}{360°} \times 2\pi r$

LEARN IT!

You can give answers in terms of π.
There is more about this on the previous page.

Worked example

The diagram shows a minor sector of a circle of radius 13 cm.

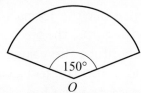

Work out the perimeter of the sector. **(4 marks)**

Arc length $= \dfrac{x}{360°} \times 2\pi r$

$= \dfrac{150°}{360°} \times 2\pi \times 13$

$= 34.033\,92\ldots$

Perimeter $=$ arc length $+$ radius $+$ radius

$= 34.033\,92\ldots + 13 + 13$

$= 60$ cm (2 s.f.)

Don't round until your final answer. The radius is given correct to 2 significant figures so this is a good degree of accuracy.

Finding a missing angle

You can use the formulae for arc length or area to find a missing angle in a sector. Practise this method to help you tackle the hardest questions.

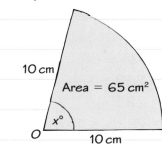

Area of sector $= \dfrac{x}{360} \times \pi r^2$

$65 = \dfrac{x}{360} \times \pi(10)^2$

$x = \dfrac{65 \times 360}{\pi(10)^2}$

$= 74.4845\ldots$

$= 74.5°$ (to 3 s.f.)

Now try this

OAB is a sector of a circle, centre O.
Angle $AOB = 60°$.
$OA = OB = 12$ cm.
Work out the length of the arc AB.
Give your answer correct to 3 significant figures. **(3 marks)**

You need to learn the formula for arc length.

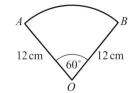

Pythagoras' theorem

Pythagoras' theorem is a really useful rule. You can use it to find the length of a missing side in a right-angled triangle.

LEARN IT!

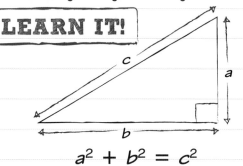

$$a^2 + b^2 = c^2$$

Pythagoras checklist

- ☑ $short^2 + short^2 = long^2$
- ☑ Side c is the hypotenuse.
- ☑ Right-angled triangle.
- ☑ Lengths of two sides known.
- ☑ Length of third side missing.
- ☑ Learn this.

Worked example

This right-angled triangle has sides x, 17 cm and 8 cm.

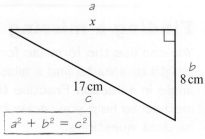

$$a^2 + b^2 = c^2$$

Show that $x = 15$ cm   **(2 marks)**

$$x^2 + 8^2 = 17^2$$
$$x^2 = 17^2 - 8^2$$
$$= 289 - 64$$
$$= 225$$
$$x = \sqrt{225} = 15 \text{ cm}$$

The question says 'Show that' so you have to show **all** your working. Be careful when the missing length is one of the **shorter** sides.

1. Label the longest side of the triangle c.
2. Label the other two sides a and b.
3. Write out the formula for Pythagoras' theorem.
4. Substitute the values for a, b and c into the formula.
5. Rearrange the formula and solve. Make sure you show **every step** in your working.
6. Write units in your answer.

Pythagoras questions come in lots of different forms. Just look for the right-angled triangle.

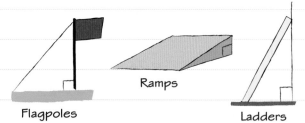

Flagpoles   Ramps   Ladders

Calculator skills

Use these buttons to find squares and square roots with your calculator.

You might need to use the S⇔D key to get your answer as a decimal number.

Now try this

(a) Work out the value of y.   **(2 marks)**

(b) Use your value of y to work out the value of z.   **(2 marks)**

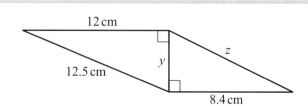

Trigonometry 1

You can use the trigonometric ratios to find the size of an angle in a right-angled triangle. You need to know the lengths of two sides of the triangle.

The sides of the triangle are labelled relative to the **angle** you need to find.

Trigonometric ratios LEARN IT!

$\sin x° = \dfrac{opp}{hyp}$ (remember this as S^O∕$_H$)

$\cos x° = \dfrac{adj}{hyp}$ (remember this as C^A∕$_H$)

$\tan x° = \dfrac{opp}{adj}$ (remember this as T^O∕$_A$)

You can use S^O$_H$C^A$_H$T^O$_A$ to remember these rules for trig ratios.

These rules only work for **right-angled** triangles.

Worked example

Calculate the size of angle x.   **(3 marks)**

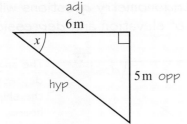

$\tan x° = \dfrac{opp}{adj} = \dfrac{5}{6}$

$x° = 39.805\,571\,09° = 39.8°$ (to 3 s.f.)

Label the **hyp**otenuse first — it's the longest side.
Then label the side **adj**acent to the angle you want to work out.
Finally, label the side **opp**osite the angle you want to work out.
Remember S^O$_H$C^A$_H$T^O$_A$. You know **opp** and **adj** here so use T^O$_A$.
Do **not** 'divide by tan' to get x on its own. You need to use the tan^{-1} function on your calculator.

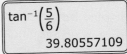

Write down all the figures on your calculator display then round your answer.

Using your calculator

To find a missing angle using trigonometry you have to use one of these functions.

$$\sin^{-1} \qquad \cos^{-1} \qquad \tan^{-1}$$

These are called **inverse trigonometric** functions. They are the inverse operations of sin, cos and tan.

Make sure that your calculator is in degree mode. Look for the ▪D symbol at the top of the display.

Now try this

Work out the size of angle x in each of these triangles. Give your answers correct to 1 decimal place.

(a)

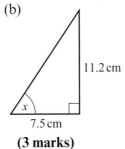

(3 marks)

(b)

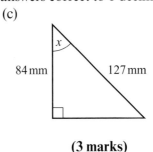

(3 marks)

(c)

(3 marks)

Trigonometry 2

You can use the trigonometric ratios to find the length of a missing side in a right-angled triangle. You need to know the length of another side and the size of one of the acute angles.

Worked example

Calculate the length of side a.   **(3 marks)**

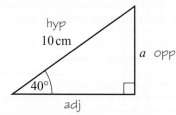

$$\sin x° = \frac{\text{opp}}{\text{hyp}}$$

$$\sin 40° = \frac{a}{10}$$

$$a = 10 \times \sin 40°$$

$$= 6.427\,87\ldots$$

$$= 6.43 \text{ cm (to 3 s.f.)}$$

Label the sides of the triangle relative to the 40° angle. Write $S^O_H C^A_H T^O_A$ and tick the pieces of information you have. You need to use S^O_H here.

Write the values you know in the rule and replace **opp** with a. You can solve this equation to find the value of a.

Write down at least four figures of the calculator display before giving your final answer correct to 3 significant figures.

Check it!

Side a must be shorter than the hypotenuse. 6.43 cm looks about right. ✓

Angles of elevation and depression

Some trigonometry questions will involve angles of elevation and depression.

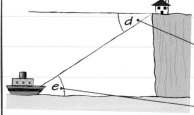

The angle of depression of the ship from the house.

The angle of elevation of the house from the ship.

Angles of elevation and depression are always measured from the horizontal.

In this diagram, $d = e$ because they are alternate angles.

Now try this

In part (c), a is the hypotenuse. It will be on the bottom of the fraction when you substitute, so be careful with your calculation.

Work out the length of side a in each of these triangles.
Give your answers correct to 1 decimal place.

(a)
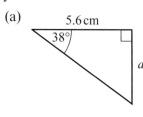

(3 marks)

(b)
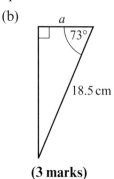

(3 marks)

(c)
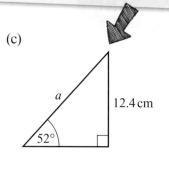

(3 marks)

Volumes of 3D shapes

You can use formulae to solve problems involving volumes of 3D shapes.

Cone

Volume of cone

$= \frac{1}{3} \times$ area of base $\times$ vertical height

$= \frac{1}{3}\pi r^2 h$

Given on the formula sheet ✓

Sphere

Volume of sphere

$= \frac{4}{3}\pi r^3$

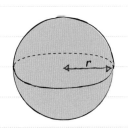

Given on the formula sheet ✓

Cuboid

Volume of cuboid

$=$ length $\times$ width $\times$ height

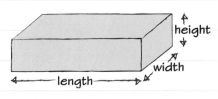

Learn this formula ✓

Worked example

The diagram shows a cone **A** and a cylinder **B**.
Show that the volume of **B** is 8 times the volume of A. **(4 marks)**

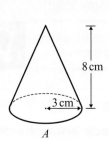 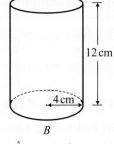

Volume of A $= \frac{1}{3}\pi r^2 h = \frac{1}{3}\pi \times 3^2 \times 8$
$\qquad\qquad = 24\pi$

Volume of B $= \pi r^2 h = \pi \times 4^2 \times 12$
$\qquad\qquad = 192\pi$

$8 \times 24\pi = 192\pi$ so the volume of B is 8 times the volume of A.

Problem solved!

You might have to **compare** two volumes or areas in your exam.
These questions might involve:
- working out the ratio between two different areas or volumes
- finding an unknown quantity represented by a letter
- finding an expression for a length, area or volume in terms of an unknown.

In this question you need to know the ratio between the two volumes. Calculate them both, then write a short **conclusion**. Make sure you show the calculation $8 \times 24\pi = 192\pi$ in your conclusion. You can leave your working in terms of π to make it easier. There is more about this on page 75.

You will need to use problem-solving skills throughout your exam – **be prepared!**

Now try this

The diagram shows a metal rivet made by joining a hemisphere to a cylinder.

Work out the volume of metal used to make the rivet. Give your answer correct to 3 significant figures. **(4 marks)**

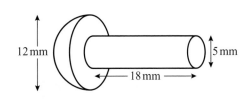

Worked solution video

The volume of a sphere is given on the formulae sheet. In this question you need to find the volume of **half** a sphere.

Surface area

Cone

The formula for the **curved surface area** of a cone is on the formulae sheet.

$$\text{Curved surface area of cone} = \pi r l$$

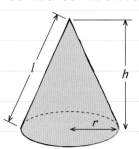

Be careful! This formula uses the slant height, l, of the cone.

To calculate the **total** surface area of the cone you need to add the area of the base. Surface area of cone $= \pi r^2 + \pi r l$

Sphere

The formula for the surface area of a sphere is on the formulae sheet.

$$\text{Surface area of sphere} = 4\pi r^2$$

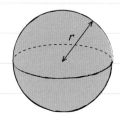

For a reminder about areas of circles and surface areas of cylinders have a look at page 74.

A hemisphere is half a sphere, so the area of the curved surface is $\frac{1}{2} \times 4\pi r^2$.

Problem solved!

To work out the curved surface area you need to know the radius and the slant height. You are given the radius and the **vertical height**. To calculate the slant height you need to use Pythagoras' theorem. Sketch the right-angled triangle containing the missing length.

You will need to use problem-solving skills throughout your exam – **be prepared!**

Worked example

The diagram shows a cone with vertical height 12 cm and base radius 5 cm. Work out the curved surface area of the cone. **(4 marks)**

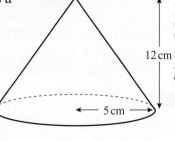

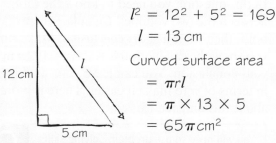

$l^2 = 12^2 + 5^2 = 169$

$l = 13$ cm

Curved surface area

$= \pi r l$

$= \pi \times 13 \times 5$

$= 65\pi$ cm^2

Compound shapes

You can calculate the surface area of more complicated shapes by adding together the external surface area of each part.

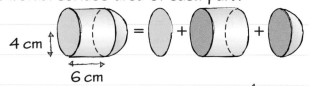

Surface area $= \pi(4)^2 + 2\pi(4)(6) + \frac{1}{2}[4\pi(4)^2]$

$= 96\pi$ cm^2

Now try this

1 Work out the surface area of this prism. **(3 marks)**

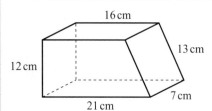

Worked solution video

Look at the blue box above for a hint.

2 A solid object is formed by joining a hemisphere to a cylinder. Both the hemisphere and the cylinder have a diameter of 4.2 cm. The cylinder has a height of 5.6 cm. Work out the total surface area of the object. Give your answer to 3 significant figures. **(4 marks)**

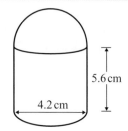

Translations, reflections and rotations

You might have to describe these transformations in your exam. To describe a translation you need to give a vector. To describe a reflection you need to give the equation of the mirror line. To describe a rotation you need to give the direction, the angle and the centre of rotation.

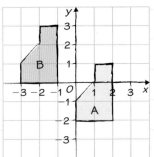

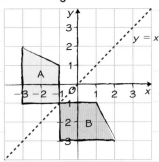

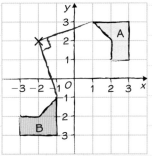

You can ask for tracing paper in an exam. This makes it easy to rotate shapes and check your answers.

A to B: Translation by the vector $\begin{pmatrix} -3 \\ 2 \end{pmatrix}$

A to B: Reflection in the line $y = x$

A to B: Rotation 90° clockwise about the point $(-2, 2)$

For all three transformations, shape **B** is **congruent** to shape **A**. This means that they are exactly the same shape and size. Lengths of sides and angles do not change.

Worked example

The diagram shows two shapes **P** and **Q**.

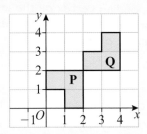

Describe fully the single transformation which takes shape **P** to shape **Q**.   **(3 marks)**

Rotation 90° clockwise with centre (3, 1).

To **fully describe** a rotation you need to write down:
- the word 'rotation'
- the angle of turn and the direction
- the centre of rotation.

Be careful when the shapes are joined at one corner. This point is not necessarily the centre of rotation.

Check it!
You are allowed to ask for tracing paper in your exam. Trace shape **P** and put your pencil on your centre of rotation. Rotate the tracing paper to see if the shapes match up. ✓

Now try this

Triangles **A**, **B** and **C** are shown on the grid.

(a) Describe fully the **single** transformation that takes triangle **A** onto triangle **B**.   **(3 marks)**

(b) Describe fully the **single** transformation that takes triangle **A** onto triangle **C**.   **(1 mark)**

For part (a) you must write a **single** transformation. You will never have to write a combined transformation in your exam.

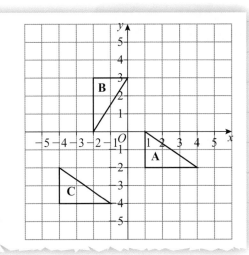

Enlargements

To describe an enlargement you need to give the scale factor and the centre of enlargement.

The **scale factor** of an enlargement tells you how much each length is multiplied by.

$$\text{Scale factor} = \frac{\text{enlarged length}}{\text{original length}}$$

Lines drawn through corresponding points on the object (A) and image (B) meet at the **centre of enlargement**.

When the scale factor is between 0 and 1, image B is **smaller** than object A.

For enlargements, angles in shapes do not change but lengths of sides do change.

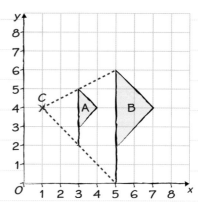

A to B: Each point on B is twice as far from C as the corresponding point on A.

Enlargement with scale factor 2, centre (1, 4)

Worked example

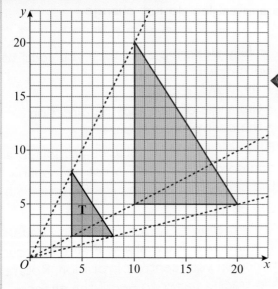

On the grid, enlarge triangle **T** with a scale factor of $2\frac{1}{2}$ and centre (0, 0)   **(3 marks)**

1. Draw lines from the centre of enlargement through each vertex of the triangle.

2. For each vertex, multiply the vertical and horizontal distances from the centre of enlargement by $2\frac{1}{2}$

   For the top vertex:

   Horizontal distance $= 4 \times 2\frac{1}{2} = 10$

   Vertical distance $= 8 \times 2\frac{1}{2} = 20$

   The corresponding vertex on the image is 10 squares horizontally and 20 squares vertically from the centre of enlargement.

3. Join up your vertices with straight lines.

Check it!

Each length on the image should be $2\frac{1}{2}$ times the corresponding length on the object. The image is **mathematically similar** to the object, so check it looks the same shape.

There is more on similar shapes on pages 87 and 88.

Now try this

Triangle **A** is shown on the grid.

(a) Enlarge triangle **A** with a scale factor of 2 and centre of enlargement (6, 5)
    Label the image **B**.   **(2 marks)**

(b) Enlarge triangle **A** with a scale factor of $\frac{1}{2}$ and centre of enlargement (−7, 3)
    Label the image **C**.   **(3 marks)**

The scale factor is a **fraction** so the image will be **smaller** than the object.

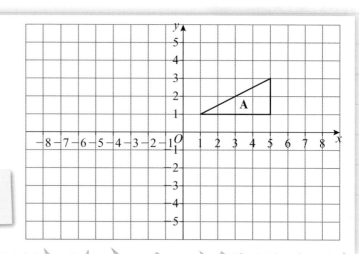

Combining transformations

You can sometimes describe two or more transformations using a single transformation.

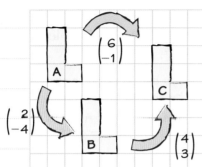

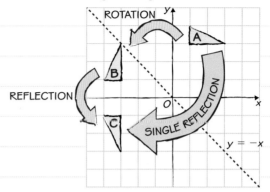

A to B to C: A translation $\begin{pmatrix} 2 \\ -4 \end{pmatrix}$ followed by a translation $\begin{pmatrix} 4 \\ 3 \end{pmatrix}$ is the same as a single translation $\begin{pmatrix} 6 \\ -1 \end{pmatrix}$.

A to B to C: A rotation 90° clockwise about O followed by a reflection in the x-axis is the same as a single reflection in the line $y = -x$.

Worked example

Triangle **A** is shown on the grid.

(a) Reflect triangle A in the y-axis. Label your new triangle **B**.
(1 mark)

(b) Triangle **B** is reflected in the line $y = 1$ to give triangle **C**. Describe fully the **single** transformation which takes triangle **A** onto triangle **C**. **(4 marks)**

Rotation 180° about the point (0, 1)

> Everything in red is part of the answer.

To answer part (b) you need to draw
• the line $y = 1$
• the new triangle, **C**.

For a rotation of 180° you don't need to give a direction.

Check it!
You can ask for tracing paper in the exam.
Check a reflection by folding the tracing paper along the symmetry line. ✓

Describe fully ...

☑ A translation: vector of translation.
☑ A reflection: equation of mirror line.
☑ A rotation: angle of turn, direction of turn and centre of rotation.
☑ An enlargement: scale factor and centre of enlargement.

Now try this

Joelle reflects a shape in the line $x = 4$.
She then reflects the image in the line $y = 5$.

Describe fully a **single** transformation that has the same effect as Joelle's two transformations. You can use this graph paper to help you. **(4 marks)**

Draw a triangle on the grid and apply the two transformations to help you see what is going on. You can use a second shape to check your answer. Make sure you write a **single** transformation in your answer.

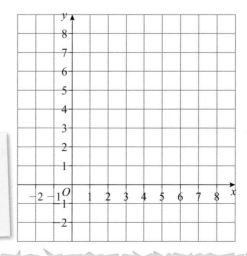

Bearings

Bearings (i.e. the angle measured from a fixed point using a compass) are measured **clockwise** from **north**.

Bearings always have **three figures**, so you need to add zeros if the angle is less than 100°. For instance, in this diagram the bearing of B from A is 048°.

You can measure a bearing bigger than 180° by measuring this angle and subtracting it from 360°.

The bearing of C from A is 360° − 109° = 251°

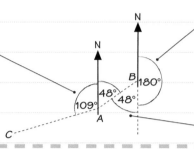

You can work out a reverse bearing by adding or subtracting 180°.

The bearing of A from B is 180° + 048° = 228°

These are alternate angles.

Worked example

The diagram shows the location of a boat on a reservoir. The reservoir has a water inlet that is 700 m from the boat on a bearing of 230°.

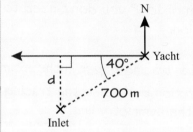

Everything in red is part of the answer.

(a) Write down the bearing of the boat from the water inlet. **(1 mark)**

230 − 180 = 50

Bearing of boat from inlet is 050°.

Boats are not allowed within 400 m of the water inlet. The boat sails due west.

(b) Show **by calculation** that the boat will not pass within 400 m of the water inlet. **(4 marks)**

$$\sin x° = \frac{opp}{hyp}$$

$$\sin 40° = \frac{d}{700} \text{ so } d = 700 × \sin 40°$$

$$= 449.951... \text{ m}$$

$d > 400$ m so the boat does not come within 400 m of the inlet.

Compass points

You need to know the compass points:

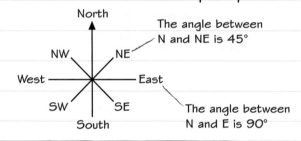

The angle between N and NE is 45°

The angle between N and E is 90°

Problem solved!

You might have to combine a bearings question with **trigonometry** calculations. You should always use a clear, well-labelled sketch to make sure you don't make any mistakes.

The shortest distance from a point to a line is a **perpendicular** line. Sketch this on the diagram and write down any lengths or angles you know. You can find the 40° angle by working out 270° − 230°.

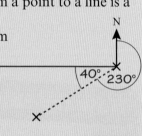

You will need to use problem-solving skills throughout your exam – **be prepared!**

Now try this

The diagram shows three orienteering markers, P, Q and R.

R is due east of Q. The bearing of Q from P is 068°.

PQ = QR.

(a) Work out the bearing of P from Q. **(2 marks)**

(b) Work out the bearing of R from P. **(3 marks)**

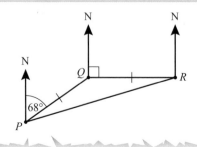

Worked solution video

Scale drawings and maps

This is a **scale drawing** of the Queen Mary II cruise ship.

Scale = 1 : 500

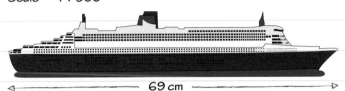

← 69 cm →

You can use the scale to work out the length of the actual ship.

69 × 500 = 34 500

The ship is 34 500 cm or 345 m long.

Map scales

Map scales can be written in different ways:

✓ 1 to 25 000

✓ 1 cm represents 25 000 cm

✓ 1 cm represents 250 m

✓ 4 cm represent 1 km

MAP SCALE 1 : 25 000

Worked example

The diagram shows a scale drawing of a port and a lighthouse.

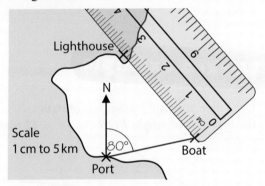

A boat sails 12 km from the port in a straight line on a bearing of 080°.

How far away is the boat from the lighthouse? Give your answer in km. **(3 mark)**

15 km

Use the scale drawing to answer the question. You need to mark the position of the boat accurately on the scale drawing.

Place the centre of your protractor on the port with the zero line pointing north. Put a dot at 80° and draw a line to show the direction of the boat from the port.

Then work out how far the boat is from the port on the scale drawing.

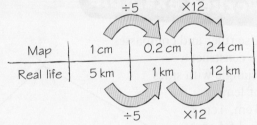

	÷5	×12	
Map	1 cm	0.2 cm	2.4 cm
Real life	5 km	1 km	12 km
	÷5	×12	

Draw a cross on your line 2.4 cm from the port.

Use a ruler to measure the distance from the lighthouse to the boat. 3 cm on the drawing represents 15 km in real life.

Now try this

1 A map has a scale of 1 : 40 000. A park is shown on the map as a rectangle measuring 5.8 cm by 4.4 cm.

Calculate the area of the park in real life. Give your answer in km² to 3 significant figures. **(3 marks)**

2 The map shows Hereford and Worcester. The scale of the map is 1 : 1 000 000. Work out the distance between Hereford and Worcester in kilometres. **(3 marks)**

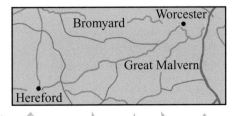

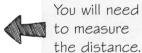

You will need to measure the distance.

Constructions

You might need to construct lines, angles or shapes using a **ruler** and **compasses**.

Worked example

Use ruler and compasses to **construct** the perpendicular bisector of the line *AB*.

(2 marks)

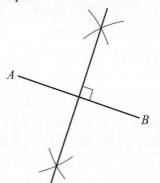

Use your compasses to draw intersecting arcs with centres at A and B. Remember that to get full marks you have to be **accurate** and show **all** your construction lines.

Everything in red is part of the answer.

Constructions checklist

✓ Use good compasses with stiff arms.
✓ Use a sharp pencil.
✓ Use a transparent ruler.
✓ Mark any angles.
✓ Label any lengths.
✓ Show all construction lines.

Worked example

Use ruler and compasses to **construct** a triangle with sides of length 3 cm, 4 cm and 5.5 cm.

(2 marks)

4 cm   3 cm   5.5 cm

Draw one side with a ruler and label it. Then use your compasses to find the other vertex.

Worked example

Use ruler and compasses to **construct** the bisector of angle *PQR*.   **(2 marks)**

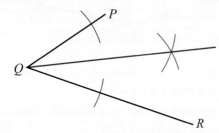

Mark points on each arm an equal distance from Q. Then use arcs to find a third point an equal distance from these two points.

Now try this

Construct the perpendicular bisector of the line *AB*.

(2 marks)

Worked solution video

Remember to show **all** your construction marks.

Similar shapes 1

Shapes are **similar** if one shape is an enlargement of the other.

Similar triangles satisfy these three conditions:

 All three pairs of angles are equal.

 All three pairs of sides are in the same ratio.

 Two sides are in the same ratio and the included angle is equal.

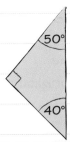 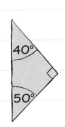

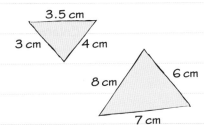

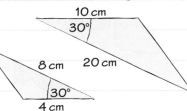

Worked example

XYZ and ABC are similar triangles.

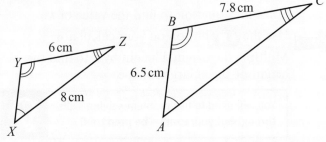

(a) Work out the length of AC. **(2 marks)**

$$\frac{AC}{XZ} = \frac{BC}{YZ}$$

$$\frac{AC}{8} = \frac{7.8}{6}$$

$$AC = \frac{7.8 \times 8}{6}$$

$$= 10.4 \, cm$$

(b) Work out the length of XY. **(2 marks)**

$$\frac{XY}{AB} = \frac{YZ}{BC}$$

$$\frac{XY}{6.5} = \frac{6}{7.8}$$

$$XY = \frac{6 \times 6.5}{7.8}$$

$$= 5 \, cm$$

Start with the unknown length on top of a fraction. Make sure you write your ratios in the correct order.

Similar shapes checklist

Use these facts to solve similar shapes problems:

✓ Corresponding angles are equal.

✓ Corresponding sides are in the same ratio.

Spotting similar triangles

Here are some similar triangles:

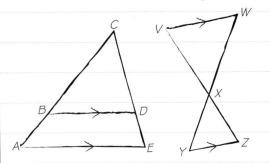

Triangle ACE is similar to triangle BCD

Triangle VWX is similar to triangle ZYX

Now try this

Triangles ABC and PQR are similar.

Angle ACB = angle PRQ

(a) Work out the size of angle PRQ. **(2 marks)**

(b) Work out the length of PQ. **(2 marks)**

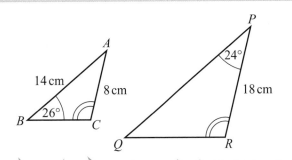

Similar shapes 2

The relationship between similar shapes is defined by a **scale factor**.
A and **B** are similar shapes. **B** is an enlargement of **A** with scale factor k.

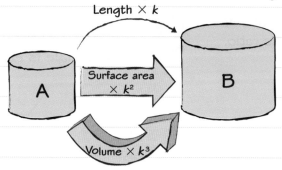

Length × k

Surface area × k^2

Volume × k^3

When a shape is enlarged by a linear scale factor k:

- Enlarged surface area
    $= k^2 ×$ original surface area
- Enlarged volume
    $= k^3 ×$ original volume
- Enlarged mass $= k^3 ×$ original mass

Worked example
Aiming higher

These two glass prisms are similar in shape.

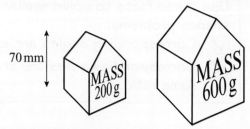

70 mm

MASS 200 g

MASS 600 g

The 200 g prism is 70 mm high.
Work out the height of the 600 g prism.

(3 marks)

$200 × k^3 = 600$

$k^3 = \dfrac{600}{200} = 3$

$k = \sqrt[3]{3} = 1.4422...$

Enlarged height $= 70 × 1.4422...$

$= 101$ mm (3 s.f.)

Problem solved!

Always use k, k^2 or k^3 to write the relationship.
Enlarged mass $= k^3 ×$ original mass
Solve the equation to find the value of k.
Use the button on your calculator.
Multiply the original height by k to find the height of the enlarged shape.

> You will need to use problem-solving skills throughout your exam – **be prepared!** 💡

Comparing volumes

You can use k^3 to compare volume, mass or capacity.

$k = \dfrac{32}{16} = 2$

Volume of large bottle

$= 1.2 × k^3$

$= 1.2 × 8$

$= 9.6$ litres

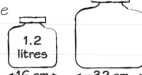

1.2 litres

◄16 cm►   ◄ 32 cm →

Now try this
Aiming higher

Here are three mathematically similar containers.
The table shows some information about these containers.

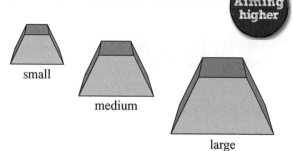

small

medium

large

	Height (cm)	Area of top of container (cm²)	Volume (cm³)
small	12	85	Y
medium	36	X	7830
large	W	2125	Z

Work out the missing values, W, X, Y and Z.

(6 marks)

The sine rule

The **sine rule** applies to any triangle. You don't need a right angle.

You label the angles of the triangle with capital letters and the sides with lower case letters. Each side has the same letter as its **opposite** angle.

$$\frac{a}{\sin A} = \frac{b}{\sin B} = \frac{c}{\sin C}$$

 **LEARN IT!**

This version is given on the formulae sheet. Use it to find the missing side.

$$\frac{\sin A}{a} = \frac{\sin B}{b} = \frac{\sin C}{c}$$

Learn this version. It's useful for finding the missing angle.

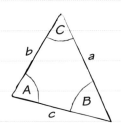

Worked example   *Aiming higher*

Calculate the size of angle x.   **(3 marks)**

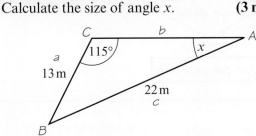

$$\frac{\sin A}{a} = \frac{\sin C}{c}$$

$$\frac{\sin x}{13} = \frac{\sin 115°}{22}$$

$$\sin x = \frac{13 \times \sin 115°}{22}$$

$$= 0.5355...$$

$$x = 32.4° \text{ (3 s.f.)}$$

> Everything in red is part of the answer.

Golden rule

To use the sine rule you need to know a side length and the **opposite** angle.

This is not a right-angled triangle so you can't use $S^O_H \, C^A_H \, T^O_A$. To find an angle, use the 'upside down' version of the sine rule. You're not interested in side b or angle B so ignore this part of the rule.

Start by writing the value you want to find on top of the first fraction. Then substitute the other values you know and solve an equation to find x. Use the $\sin^{-1}$ function on your calculator.

Check your answer makes sense. The diagram is not to scale but it should look about right.

Worked example   *Aiming higher*

Calculate the length of AC.   **(3 marks)**

$$\frac{b}{\sin B} = \frac{a}{\sin A}$$

$$\frac{AC}{\sin 70°} = \frac{6.3}{\sin 52°}$$

$$AC = \frac{6.3 \times \sin 70°}{\sin 52°}$$

$$= 7.5126...$$

$$= 7.5 \text{ cm (2 s.f.)}$$

> Everything in red is part of the answer.

You know a side length and the opposite angle so you can use the sine rule.

Check it!

The greater side length is opposite the greater angle. ✓

Now try this   *Aiming higher*

In triangle ABC, $AB = 11$ cm, $AC = 13$ cm and angle $ABC = 57°$

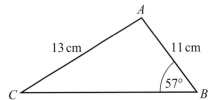

(a) Work out the size of angle ACB.   **(3 marks)**
(b) Work out the length of BC.   **(3 marks)**

89

The cosine rule

The **cosine rule** applies to any triangle. You don't need a right angle.

You usually use the cosine rule when you are given two sides and the included angle (SAS) or when you are given three sides and want to work out an angle (SSS).

$$a^2 = b^2 + c^2 - 2bc \cos A$$

This version is on the formulae sheet. Use it to find the missing side.

$$\cos A = \frac{b^2 + c^2 - a^2}{2bc}$$

Learn this version. It's useful for finding a missing angle.

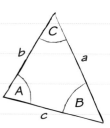

Which rule?

This chart shows you which rule to use when solving trigonometry problems in triangles:

Right-angled triangle? — **NO** → Side and the opposite angle given? — **NO** → Use the cosine rule

YES ↓ Use $S^O_H C^A_H T^O_A$

YES ↓ Use the sine rule

Worked example

Aiming higher

PQRS is a trapezium. Work out the length of the diagonal *PR*. **(3 marks)**

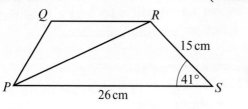

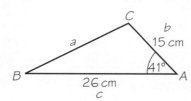

$$a^2 = b^2 + c^2 - 2bc \cos A$$

$$PR^2 = 15^2 + 26^2 - 2 \times 15 \times 26 \times \cos 41°$$

$$= 312.3265\ldots$$

$$PR = 17.6727\ldots = 17.7 \text{ cm (3 s.f.)}$$

If you are given a more complicated diagram it is sometimes useful to sketch a triangle. Label your triangle with *a* as the missing side.

This is not a right-angled triangle so you can't use $S^O_H C^A_H T^O_A$. You know two sides and the included angle (SAS) so you can use the cosine rule.

Substitute the values you know into the formula. Work out the right-hand side using your calculator, but don't round your answer yet.

Use √☐ Ans on your calculator to find the final answer.

Round to 3 significant figures or to the same degree of accuracy as the original measurements. Because this answer shows all its workings, you could give either 17.7 cm or 18 cm as the answer.

Now try this

Aiming higher

Triangle *PQR* has sides of 9 cm, 10 cm and 14 cm.

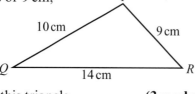

Work out the size of the smallest angle in this triangle. **(3 marks)**

Use the cosine rule if you are given **three sides** and you need to find an angle. You should use this version of the cosine rule:

$$\cos A = \frac{b^2 + c^2 - a^2}{2bc}$$

The smallest angle is always opposite the smallest side.

Triangles and segments

When you know the lengths of two sides and the angle **between them**, the area of any triangle can be found using this formula:

Area $= \frac{1}{2} ab \sin C$

You can use this formula for **any** triangle. You don't need to have a right angle.

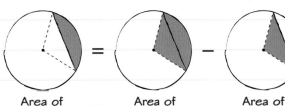

Areas of segments

A chord divides a circle into two **segments**.

Area of minor segment = Area of whole sector − Area of triangle

---

Worked example   *Aiming higher*

The diagram shows a sector of a circle with centre O.

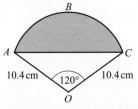

Work out the area of the shaded segment ABC. Give your answer correct to 3 significant figures.   **(5 marks)**

Whole sector $OABC$:
Area $= \dfrac{120}{360} \times \pi \times 10.4^2$

  $= 113.2648\ldots$ cm²

Triangle OAC:
Area $= \frac{1}{2} \times 10.4 \times 10.4 \times \sin 120°$
  $= 46.8346\ldots$ cm²

Shaded segment ABC:
Area $= 113.2648\ldots - 46.8346\ldots$
  $= 66.4302\ldots$
  $= 66.4$ cm² (to 3 s.f.)

If you are aiming for a top grade you need to be able to calculate the area of a sector and a triangle.

To get full marks you need to keep track of your working. Make sure you write down exactly what you are calculating at each step.

Remember that 10.4 cm is the length of one side of the triangle **and** the radius of the circle.

Make sure you don't round too soon. Write down all the figures from your calculator display at each step. Only round your **final answer** to 3 significant figures.

Which formula?

If you know the base and the vertical height:

Area $= \frac{1}{2} \times$ base $\times$ vertical height
  $= \frac{1}{2} \times 6 \times 2.1$
  $= 6.3$ cm²

If you know two sides and the included angle:

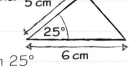

Area $= \frac{1}{2} ab \sin C$
  $= \frac{1}{2} \times 5 \times 6 \times \sin 25°$
  $= 6.3\ldots$ cm²

Now try this   *Aiming higher*

The diagram shows a circle, radius 4 cm, with angle $AOB = 135°$.
Using the fact that $\sin 135° = \sin 45°$, show clearly that the shaded area is equal to $6\pi - 4\sqrt{2}$ cm².   **(5 marks)**

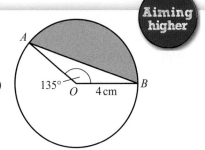

Pythagoras in 3D

To tackle the most demanding questions, you need to be able to use Pythagoras' theorem in 3D shapes.

You can use Pythagoras' theorem to find the length of the longest diagonal in a cuboid.

You can also use Pythagoras to find missing lengths in pyramids and cones.

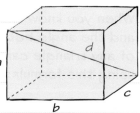

$$a^2 + b^2 + c^2 = d^2$$

Why does it work?

You can use 2D Pythagoras twice to show why the formula for 3D Pythagoras works.

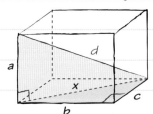

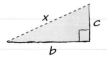

$$x^2 = b^2 + c^2$$

$$d^2 = a^2 + x^2$$
$$= a^2 + b^2 + c^2$$

Worked example

Aiming higher

The diagram shows a cuboid. Work out the length of *PQ*.
(3 marks)

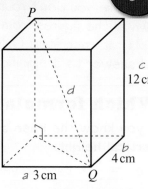

$$d^2 = a^2 + b^2 + c^2$$
$$PQ^2 = 3^2 + 4^2 + 12^2$$
$$= 169$$
$$PQ = \sqrt{169} = 13$$

So *PQ* is 13 cm.

> Everything in red is part of the answer.

Write out the formula for Pythagoras in 3D. Label the sides of the cuboid *a*, *b* and *c*, and label the long diagonal *d*.

You could also answer this question by sketching two right-angled triangles and using 2D Pythagoras.

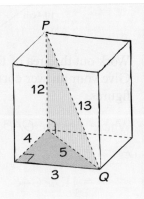

Check it!

The diagonal must be longer than any of the other three lengths.

13 cm looks about right. ✓

Now try this

Aiming higher

The diagram shows Bridget's new sewing box and a knitting needle.

Will the knitting needle fit inside the box?

You must show all of your working.   **(3 marks)**

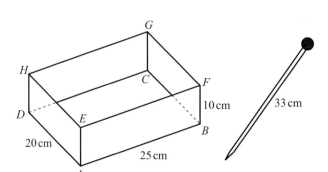

Worked solution video

Trigonometry in 3D

You can use $S^O_H C^A_H T^O_A$ to find the angle between a **line** and a **plane**.

You might need to combine trigonometry and Pythagoras' theorem when you are solving 3D problems.

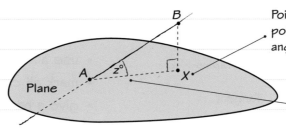

Point X is directly below point B, so ABX is a right-angled triangle.

Angle z is the angle between the line and the plane.

Worked example

Aiming higher

The diagram shows a triangular prism.

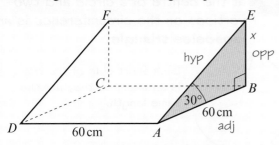

Calculate the angle between the line *DE* and the base of the prism. **(6 marks)**

△ABE: you know one angle and the adjacent side. You are looking for the opposite side, so use T^O_A.

△DAB: you know two sides so you can use Pythagoras' theorem.

△DEB: you know the opposite and adjacent sides so use T^O_A. Use $\tan^{-1}$ to find the value of z.

Do **not** round any of your answers until the end — write down at least six figures from each calculator display.

$\tan 30° = \dfrac{x}{60}$

$x = 60 \times \tan 30°$

$\quad = 34.6410\ldots \text{cm}$

Everything in red is part of the answer.

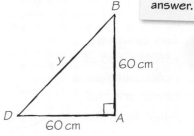

$y^2 = 60^2 + 60^2$

$\quad = 7200$

$y = \sqrt{7200} = 84.8528\ldots \text{cm}$

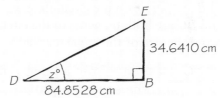

$\tan z° = \dfrac{34.6410\ldots}{84.8528\ldots} = 0.40824\ldots$

$z° = \tan^{-1} 0.408\,24\ldots = 22.2076\ldots$

$\quad = 22.2°$ (to 3 s.f.)

Now try this

Aiming higher

The diagram shows a triangular-based pyramid.
OAC and *OCB* are right-angled triangles.
Work out the size of angle *BOC*.
Give your answer correct to 1 decimal place. **(6 marks)**

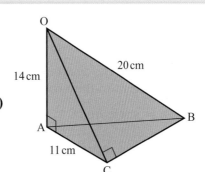

Worked solution video

Start by sketching triangle OAC and working out the length of OC.

93

Circle facts

You need to know the names of the different parts of a circle.

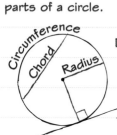

Diameter = radius × 2

Tangent

The other parts of a circle are shown on pages 74 and 75.

When you are solving circle problems:

- correctly identify the angle to be found
- use all the information given in the question
- mark all calculated angles on the diagram
- give a reason for each step of your working.

You might need to use angle facts about triangles, quadrilaterals and parallel lines in circle questions. There is a list of angle facts on page 68.

Key circle facts

 The angle between a radius and a tangent is 90°.

 Two tangents which meet at a point outside a circle are the same length.

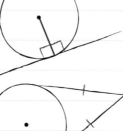

3 A triangle which has one vertex at the centre of a circle and two vertices on the circumference is an **isosceles triangle**.

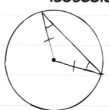

Each short side of the triangle is a radius, so they are the same length.

Remember that the base angles of an isosceles triangle are equal.

Worked example

A and B are points on the circumference of a circle with centre O. AC and BC are both tangents to the circle. Angle $BCA = 42°$
Work out the size of the angle marked x.   **(3 marks)**

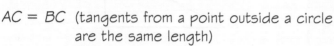

$AC = BC$ (tangents from a point outside a circle are the same length)

$\angle ABC = \dfrac{180° - 42°}{2} = 69°$

(base angles in an isosceles $\triangle$ are equal, and angles in a $\triangle$ add up to 180°)

$x + 69° = 90°$ (angle between a tangent and a radius = 90°)

$x = 21°$

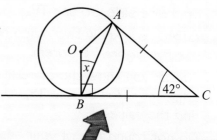

$AC = BC$, so mark these lines with a dash. Make sure you also write down the circle fact you are using. To write a really good answer you have to give a reason for each step of your working.

Now try this

Aiming higher

A, B and C are points on the circumference of a circle with centre O. CD is a tangent to the circle. Angle $AOB = 53°$

Work out the size of angle BCD.   **(3 marks)**

Make sure you write down every step of your working.

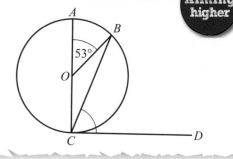

Intersecting chords

You need to know these rules about intersecting chords in the same circle.

 Inside the circle

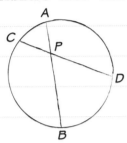

$AP \times PB = CP \times PD$

 Outside the circle

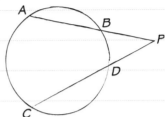

$AP \times PB = CP \times PD$

These two rules are the same — note that all four line segments used in the formula end at the point of intersection, P. These rules are not on the formula sheet, so **learn** them.

Worked example

Aiming higher

PTR and *QTS* are chords of a circle.

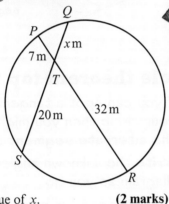

$PT = 7\,\text{m}$
$ST = 20\,\text{m}$
$RT = 32\,\text{m}$
$QT = x\,\text{m}$

Calculate the value of x.   **(2 marks)**

$ST \times QT = PT \times RT$
$20 \times x = 7 \times 32$
$20x = 224 \quad (\div 20)$
$x = 11.2$

Write down the formula before you substitute any values. Then substitute the values you are given and solve the equation to find x.

Check it!
Make sure you are multiplying line segments on the **same** line:
ST and QT ✓   PT and RT ✓

Chord and tangent

If a chord intersects with a tangent, then you can use this special case:

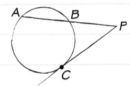

$AP \times BP = CP^2$

Now try this

Aiming higher

AB and *CD* are chords of a circle.
EP is a tangent to the circle.
$AB = 2\,\text{cm}$, $BP = 3\,\text{cm}$ and $DP = 2.5\,\text{cm}$
$CD = x\,\text{cm}$ and $EP = y\,\text{cm}$

(a) Calculate the value of x.   **(3 marks)**

(b) Calculate the value of y, giving your answer correct to 1 decimal place.   **(2 marks)**

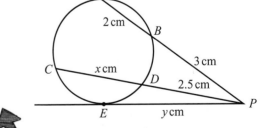

Be really careful with part (a). You will be using the formula $AP \times BP = CP \times DP$ so you need to use the following lengths:
$AP = 2 + 3 = 5\,\text{cm}$ $BP = 3\,\text{cm}$ $DP = 2.5\,\text{cm}$
The formula will tell you the length of CP, so you need to subtract 2.5 to find x.

Circle theorems

You need to **learn** these six circle theorems and know how to apply them to solve problems.

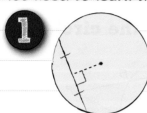 The perpendicular from a chord to the centre of the circle bisects the chord.

LEARN IT!

 Opposite angles of a cyclic quadrilateral add up to 180°.

 The angle in a semicircle is 90°.

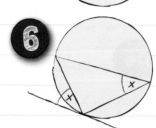 Angles in the same segment are equal.

 The angle at the centre of the circle is twice the angle on the circumference.

The angle between a tangent and a chord is equal to the angle in the alternate segment.

This is called the **alternate segment theorem**.

Worked example

Aiming higher

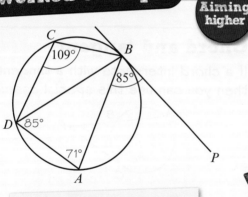

Everything in red is part of the answer.

Angle $ABP = 85°$ and angle $BCD = 109°$
Calculate the size of angle ABD.
Give reasons for each step of your working. **(4 marks)**

$\angle ADB = 85°$ (alternate segment theorem)

$180 - 109 = 71$

$\angle DAB = 71°$ (opposite angles in a cyclic quadrilateral add up to 180°)

$180 - 85 - 71 = 24$

$\angle ABD = 24°$ (angles in a triangle add up to 180°)

Circle theorem top tips

✓ If you can spot a tangent and a chord in your circle then you might be able to use the **alternate segment theorem**.

✓ **Write** the unknown angles on your diagram as you go.

✓ Give **reasons** for each step of your working.

You might need to use other angle facts in a circle theorem question. Look at page 68 for a reminder.

Now try this

Aiming higher

$ABCD$ is a cyclic quadrilateral.
PBQ is a tangent to the circle at B.

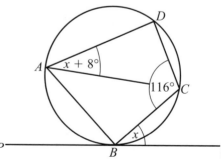

Work out the value of x. **(4 marks)**

Vectors

A vector has a **magnitude** (or size) and a **direction**.

This vector can be written as $\mathbf{a}$, $\overrightarrow{AB}$ or $\begin{pmatrix} 2 \\ 5 \end{pmatrix}$.

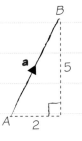

You can multiply a vector by a number. The new vector has a different length but the same direction.

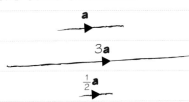

If $\mathbf{b}$ is a vector then $-\mathbf{b}$ is a vector with the same length but opposite direction.

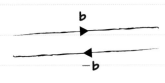

Worked example

In the diagram, $OADB$ and $ACED$ are two identical parallelograms.

$\mathbf{a}$ is the vector $\overrightarrow{OA}$

$\mathbf{b}$ is the vector $\overrightarrow{OB}$

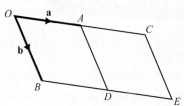

Find the following vectors in terms of $\mathbf{a}$ and $\mathbf{b}$.

(a) $\overrightarrow{OD}$ **(1 mark)**

$\overrightarrow{OD} = \overrightarrow{OA} + \overrightarrow{AD}$

$\quad = \mathbf{a} + \mathbf{b}$

(b) $\overrightarrow{EB}$ **(1 mark)**

$\overrightarrow{EB} = \overrightarrow{ED} + \overrightarrow{DB}$

$\quad = -\mathbf{a} + -\mathbf{a} = -2\mathbf{a}$

(c) $\overrightarrow{BC}$ **(1 mark)**

$\overrightarrow{BC} = \overrightarrow{BD} + \overrightarrow{DE} + \overrightarrow{EC}$

$\quad = \mathbf{a} + \mathbf{a} + -\mathbf{b} = 2\mathbf{a} - \mathbf{b}$

Adding vectors

You can add vectors using the **triangle law**. You trace a path along the added vectors to find the new vector.

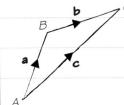

$\mathbf{a} + \mathbf{b} = \mathbf{c}$

$\mathbf{c}$ is the resultant vector of $\mathbf{a}$ and $\mathbf{b}$.

If $\mathbf{a} = \begin{pmatrix} 2 \\ 4 \end{pmatrix}$ and $\mathbf{b} = \begin{pmatrix} 6 \\ 3 \end{pmatrix}$

then $\mathbf{c} = \begin{pmatrix} 2 + 6 \\ 4 + 3 \end{pmatrix} = \begin{pmatrix} 8 \\ 7 \end{pmatrix}$

For each vector, trace a path along the shape from the start point to the end point. If you go in the **opposite** direction to the vector then you need to **subtract**.
$\overrightarrow{AD} = \overrightarrow{OB}$ because they are **parallel**.

Simplify your vectors as much as possible.

Magnitude

You can find the magnitude of a vector using Pythagoras' theorem. You use vertical lines to show the magnitude of a vector:

$\left| \overrightarrow{AB} \right| = \left| \begin{pmatrix} 2 \\ -4 \end{pmatrix} \right| = |2\mathbf{i} - 4\mathbf{j}|$

$\quad = \sqrt{2^2 + 4^2} = 2\sqrt{5}$

Ignore minus signs when calculating the magnitude of a vector.

☑ **unit vectors** have magnitude 1.

☑ The **distance** between two points A and B is the magnitude of vector $\overrightarrow{AB}$

Now try this

1   The vectors $\mathbf{a}$ and $\mathbf{b}$ are defined as

$\mathbf{a} = \begin{pmatrix} 3 \\ 5 \end{pmatrix}$ $\qquad$ $\mathbf{b} = \begin{pmatrix} 2 \\ -9 \end{pmatrix}$

(a) Write as a column vector

(i) $2\mathbf{a}$ **(1 mark)**

(ii) $\mathbf{a} + \mathbf{b}$ **(1 mark)**

(iii) $\mathbf{b} - 3\mathbf{a}$ **(3 marks)**

(b) Calculate $|\mathbf{a}|$. Give your answer correct to 1 decimal place. **(3 marks)**

Vector proof

Parallel vectors

If one vector can be written as a **multiple** of the other then the vectors are **parallel**.

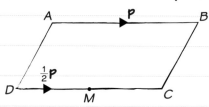

In this parallelogram M is the midpoint of DC.
AB is parallel to DM so $\overrightarrow{DM} = \frac{1}{2}\overrightarrow{AB}$

Remember that AB means the line segment AB (or the length of the line segment AB). $\overrightarrow{AB}$ means the vector which takes you from A to B.

Collinear points

If three points lie on the **same straight line** then they are collinear. Here are three points:

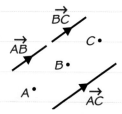

If any **two** of the vectors $\overrightarrow{AB}$, $\overrightarrow{BC}$ or $\overrightarrow{AC}$ are parallel, then the three points must be collinear.

Worked example · *Aiming higher*

In the diagram, OPQ is a triangle. Point R lies on the line PQ such that $PR:RQ = 1:2$
Point S lies on the line through OR such that $OR:OS = 1:3$

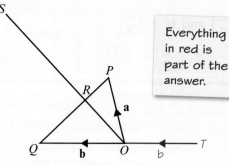

> Everything in red is part of the answer.

(a) Show that $\overrightarrow{OS} = 2\mathbf{a} + \mathbf{b}$ **(3 marks)**

$\overrightarrow{PQ} = -\mathbf{a} + \mathbf{b}$, so $\overrightarrow{PR} = \frac{1}{3}(-\mathbf{a} + \mathbf{b})$

$\overrightarrow{OR} = \mathbf{a} + \frac{1}{3}(-\mathbf{a} + \mathbf{b})$

$\quad = \frac{1}{3}(2\mathbf{a} + \mathbf{b})$

$\overrightarrow{OS} = 3\overrightarrow{OR}$

$\quad = 2\mathbf{a} + \mathbf{b}$

(b) Point T is added to the diagram such that $\overrightarrow{TO} = \mathbf{b}$. Prove that points T, P and S lie on the same straight line. **(3 marks)**

$\overrightarrow{TP} = \mathbf{a} + \mathbf{b}$

$\overrightarrow{TS} = \mathbf{b} + 2\mathbf{a} + \mathbf{b}$

$\quad = 2\mathbf{a} + 2\mathbf{b}$

$\quad = 2(\mathbf{a} + \mathbf{b})$

$\overrightarrow{TS} = 2\overrightarrow{TP}$ so $\overrightarrow{TS}$ and $\overrightarrow{TP}$ are parallel. Both vectors pass through T so they lie on the same straight line.

Problem solved!

You might be given information about the lengths of lines as ratios.

$PR:RQ = 1:2$ — There are $2 + 1 = 3$ parts in this ratio. This means that R is $\frac{1}{3}$ of the way along PQ so $\overrightarrow{PR} = \frac{1}{3}\overrightarrow{PQ}$

$OR:OS = 1:3$ — This means that $\overrightarrow{OS} = 3\overrightarrow{OR}$

To show that T, P and S lie on the same straight line you need to show that **two** of the vectors $\overrightarrow{TP}$, $\overrightarrow{TS}$ and $\overrightarrow{PS}$ are parallel.

> You will need to use problem-solving skills throughout your exam – **be prepared!**

Now try this · *Aiming higher*

In triangle OAB, M is the midpoint of OA.
$OH = HJ = JK = KB$
S and T divide the line AB into three equal segments.
$\overrightarrow{OM} = \mathbf{a}$ and $\overrightarrow{OH} = \mathbf{b}$

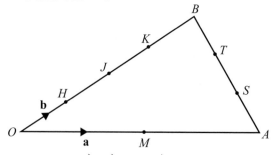

(a) Prove that $\overrightarrow{HA}$, $\overrightarrow{JS}$ and $\overrightarrow{KT}$ are all parallel. **(6 marks)**

(b) State the ratio $KT:JS:HA$ **(1 mark)**

Problem-solving practice 1

In your International GCSE Maths exams you will need to demonstrate **problem-solving** and **reasoning skills**. If you come across a tricky or unfamiliar question in your exam, you can try some of these strategies:

- ☑ Sketch a diagram to see what is going on.
- ☑ Try the problem with smaller or easier numbers.
- ☑ Plan your strategy before you start.
- ☑ Write down any formulae you might be able to use, or check the formulae sheet for a clue.
- ☑ Use x or n to represent an unknown value.

 A ladder is 6 m long.

The ladder is placed on horizontal ground, resting against a vertical wall.

The instructions for using the ladder say that the bottom of the ladder must not be closer than 1.5 m to the bottom of the wall.

How far up the wall can the ladder reach if the instructions are followed?   **(3 marks)**

Pythagoras' theorem page 76   *Aiming higher*

You should definitely draw a sketch to show the information in the question.

TOP TIP

Be careful when you are working out the length of a **short** side using Pythagoras' theorem.

Remember: $\text{short}^2 + \text{short}^2 = \text{long}^2$

$\text{short}^2 = \text{long}^2 - \text{short}^2$

 The diagram shows four metal ball bearings of diameter 2.7 cm packed inside the smallest possible box in the shape of a cuboid.

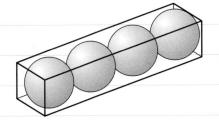

Worked solution video

Volume of a sphere = $\frac{4}{3}\pi r^3$

Work out the volume of empty space inside the box. Give your answer correct to 3 significant figures.   **(4 marks)**

Volumes of 3D shapes page 79   *Aiming higher*

You need to work out the dimensions of the box. You can use the diagram to help – once you have worked out each dimension, write it on the diagram.

Don't round any values until your final answer. You can use the ANS button on your calculator to enter the exact answer to a previous calculation. But don't write 'ANS' in your working. It's better to write down the actual number with at least 4 decimal places.

TOP TIP

Don't forget to use the formulae sheet. The formula for the volume of a sphere is given in your exam.

Problem-solving practice 2

3

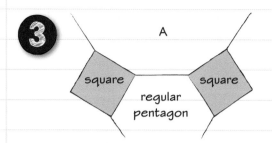

A
square   square
regular pentagon

The diagram shows two squares, part of a regular pentagon and part of a regular *n*-sided polygon, A.

Calculate the value of *n*. Show your working clearly. **(5 marks)**

Angles in polygons page 70

Follow these steps:

1. Work out the interior angles of a regular pentagon and a square.
2. Use the fact that the angles around a point add up to 360° to find the interior angle of A.
3. Subtract this from 180° to find the exterior angle of A.
4. Divide 360° by this to work out *n*.

TOP TIP

If there are a lot of steps in a question it's a good idea to **plan** your answer before you start.

4

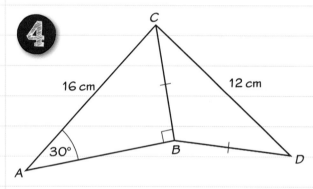

C
16 cm   12 cm
30°   B
A   D

$AC = 16\,cm$ $CD = 12\,cm$ $BC = BD$

Angle $ABC = 90°$

Angle $CAB = 30°$

Work out the area of triangle BCD.

(6 marks)

The cosine rule page 90 · *Aiming higher*
Triangles and segments page 91

You need more information about triangle BCD before you can calculate its area.

Triangle ABC is right-angled so you can use $S^O_HC^A_HT^O_A$ to work out the length of BC. Then use the cosine rule to work out the size of angle BCD. Finally, you can use the formula

$A = \frac{1}{2}ab \sin C$ to work out the area of triangle BCD.

TOP TIP

The diagram is **not drawn accurately**, so you can't measure lengths or angles.

5

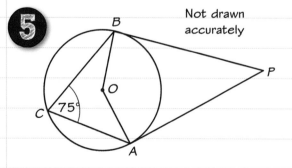

B   Not drawn accurately
O   P
C 75°
A

In the diagram A, B and C are points on the circumference of a circle with centre O.

PA and PB are tangents to the circle.

Angle $ACB = 75°$

Work out the size of angle APB.
Show your working clearly.   **(4 marks)**

Circle theorems p. 97 · *Aiming higher*
Circle facts p. 95

You have to write down a reason for **each step** of your working. These are some of the reasons you could use to answer this question:

- Angle at the centre of a circle is twice the angle at the circumference.
- Angle between a tangent and a radius is 90°.
- Angles in a quadrilateral add up to 360°.

TOP TIP

When you are learning circle theorems, draw a sketch to explain each one. This will help you to spot which theorem to use in an exam question.

Mean, median and mode

You can analyse data by calculating statistics like the **mean**, **median** and **mode**.

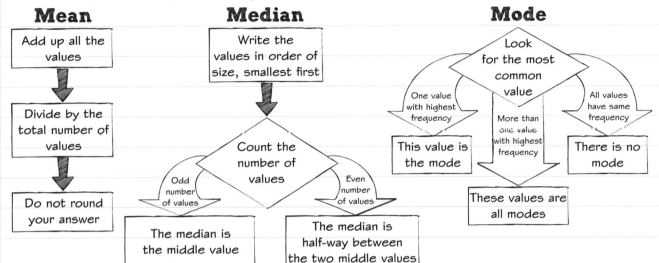

Worked example

Kayla has eight numbered cards.

She removes two cards. The mean value of the remaining cards is 4. Which two cards could Kayla have removed? Give **one** possible answer.

(4 marks)

$6 \times 4 = 24$

$1 + 2 + 3 + 4 + 5 + 6 + 7 + 8$
$= 36$

$36 - 24 = 12$

The removed cards add up to 12 so Kayla could have removed 7 and 5

Check:

$$\frac{1 + 2 + 3 + 4 + 6 + 8}{6} = 4 ✔$$

You can work out the sum of the 6 remaining cards using this formula:

Sum of values = mean × number of values

Subtract this sum from the sum of all 8 cards. This tells you the sum of the 2 cards Kayla removed. The removed cards were either 5 and 7 or 8 and 4.

Check it!

Work out the mean of the remaining 6 cards.

Which average works best?

	👍	👎
Mean	Uses all the data	Affected by extreme values
Median	Not affected by extreme values	May not be one of the values
Mode	Suitable for data that can be described in words	Not always near the middle of the data

Now try this

Takeshi scored these marks out of 20 in **six** maths tests.

    11   9   5   13   15   12

How many marks must he score in the next test so that his new mean mark and his new median mark are the same as each other?

Make sure you check your answer by calculating the new mean and median. Remember that the median is not affected by extreme values.

(3 marks)

Frequency table averages

This page shows you how to find averages from data given in frequency tables. Have a look at pages 104 and 105 to revise finding averages from graphs.

This frequency table shows the numbers of pets owned by the students in a class.

The mode is 1. This value has the highest frequency.

Number of pets (x)	Frequency (f)	Frequency × number of pets (f × x)
0	12	12 × 0 = 0
1	18	18 × 1 = 18
2	5	5 × 2 = 10
3	2	2 × 3 = 6
Total	37	34

To calculate the mean you need to add a column for 'f × x'.

There are 37 values so the median is the $\frac{37+1}{2} = 19$th value.

The first 12 values are all 0. The next 18 values are 1. So the median is 1.

The total in the 'f × x' column represents the total number of pets owned by the class.

$$\text{Mean} = \frac{\text{total number of pets}}{\text{total frequency}} = \frac{34}{37} = 0.92 \text{ (to 2 d.p.)}$$

Worked example

Fahad recorded the times, in minutes, taken by 150 students to travel to school.
The table shows his results.

Time (t minutes)	Frequency (f)	Mid-point (x)	f × x
$0 \le t < 20$	65	10	65 × 10 = 650
$20 \le t < 30$	40	25	40 × 25 = 1000
$30 \le t < 40$	39	35	39 × 35 = 1365
$40 \le t < 60$	6	50	6 × 50 = 300

Total frequency = 150   Total of f × x = 3315

Everything in red is part of the answer.

(a) Work out an estimate for the mean number of minutes that the students took to travel to school.
(4 marks)

$$\frac{3315}{150} = 22.1$$

(b) Explain why your answer to part (a) is an estimate.
(1 mark)

Because you don't know the exact data values.

Add extra columns to the table for 'Mid-point (x)' and 'Mid-point × frequency (f × x)'.

$$\text{Estimate of mean} = \frac{\text{Total of fx column}}{\text{Total frequency}}$$

Your answer does not have to be a whole number. Write down all the digits from your calculator display then round **if necessary**.

Now try this

Emma recorded the reaction times of the members of her class using a computer program. The table shows her results.

Work out an estimate for the mean reaction time.   **(4 marks)**

Reaction time (t seconds)	Frequency
$0 \le t < 0.1$	2
$0.1 \le t < 0.2$	9
$0.2 \le t < 0.3$	16
$0.3 \le t < 0.4$	5

Worked solution video

Interquartile range

Range and interquartile range are measures of spread. They tell you how spread out data is.

Quartiles divide a data set into four equal parts.

Half of the values lie between the lower quartile and the upper quartile.

$Q_1 = \dfrac{n+1}{4}$th value, where n = number of data values

|← Interquartile range (IQR) →|

× ×   × × × ×   × × × ×   ×

Smallest value — Lower quartile (Q_1) — Median (Q_2) — Upper quartile (Q_3) — Largest value

$Q_3 = \dfrac{3(n+1)}{4}$th value

Range = largest value − smallest value
Interquartile range (IQR) = upper quartile (Q_3) − lower quartile (Q_1)

Worked example

Maya recorded the heights, in cm, of some tree saplings. She put the heights in order.

21 23 23 25 26 26 31 32
33 35 36 40 40 41 42

Work out the interquartile range of Maya's data.
(3 marks)

$n = 15$

$\dfrac{n+1}{4} = \dfrac{15+1}{4} = 4$

$Q_1 = 4$th value $= 25$ cm

$\dfrac{3(n+1)}{4} = \dfrac{3(15+1)}{4} = 12$

$Q_3 = 12$th value $= 40$ cm

$IQR = Q_3 - Q_1 = 40 - 25 = 15$ cm

To work out the interquartile range, you need to know the lower quartile and the upper quartile.

1. Count the total number of values, n.
2. Check that the data is arranged in order of size.
3. Find the $\dfrac{n+1}{4}$th data value.
   This is the lower quartile (Q_1).
4. Find the $\dfrac{3(n+1)}{4}$th data value.
   This is the upper quartile (Q_3).
5. Subtract the lower quartile from the upper quartile to find the interquartile range.

Comparing distributions

You can use averages like the **mean** or **median** and measures of spread like the **range** and **interquartile range** to compare two sets of data. Follow these steps:

 1 Calculate an average and a measure of spread for both data sets.

 2 Write a sentence for each statistic **comparing** the values for each data set.

 3 Only make a statement if you can back it up with **statistical evidence**.

Now try this

Henry measures the heights in cm of some plant seedlings for an experiment. He writes the heights in order.

40 42 44 44 45 49 51
53 53 54 57 58 59 62
62 65 66 68 71

(a) How many plant seedlings did Henry measure? **(1 mark)**

(b) Work out the median height. **(2 marks)**

(c) Work out the interquartile range of this data. **(3 marks)**

Cumulative frequency

In your exam you might have to draw a cumulative frequency graph, or use one to find the median or the interquartile range.

How to draw a cumulative frequency graph

Reaction time t (s)	Frequency	Cumulative frequency
$0 < t \leqslant 0.1$	2	2
$0.1 < t \leqslant 0.2$	5	$2 + 5 = 7$
$0.2 < t \leqslant 0.3$	18	$7 + 18 = 25$
$0.3 < t \leqslant 0.4$	5	$25 + 5 = 30$
$0.4 < t \leqslant 0.5$	1	$30 + 1 = 31$

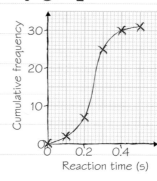

1. Plot 0 at the beginning of the first class interval.

2. Plot each value at the **upper** end of its class interval.

3. Join your points with a **smooth curve**.

Add a column for **cumulative frequency** to your frequency table.

Check that your final value is the same as the total frequency.

Here's another example:

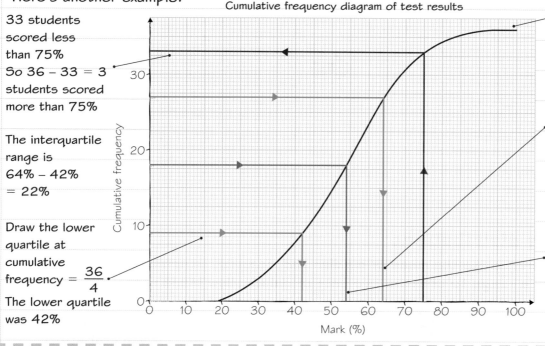

Cumulative frequency diagram of test results

33 students scored less than 75%
So 36 – 33 = 3 students scored more than 75%

There were 36 students in the class. (This is the **first fact** you should establish.)

The interquartile range is
64% – 42%
= 22%

Draw the upper quartile at cumulative frequency = $3 \times \dfrac{36}{4}$
The upper quartile was 64%

Draw the lower quartile at cumulative frequency = $\dfrac{36}{4}$
The lower quartile was 42%

Draw the median at cumulative frequency = $\dfrac{36}{2}$
The median was 54%

Now try this

This cumulative frequency graph shows the reaction times recorded by a group of students in an experiment.

Estimate the median and interquartile range of the reaction times.   **(3 marks)**

Draw lines on your graph to show the values you are reading off.

Worked solution video

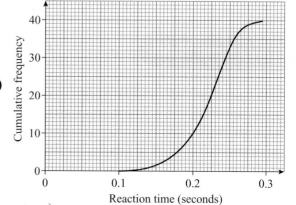

Histograms

Histograms are a good way to represent grouped data with **different** class widths.

Worked example

Aiming higher

This table shows the finishing times in minutes of runners in a cross-country race.

Time, t (minutes)	Frequency	Frequency density
$16 \leqslant t < 20$	12	3
$20 \leqslant t < 30$	45	4.5
$30 \leqslant t < 50$	28	1.4

Draw a histogram to represent the data.

(3 marks)

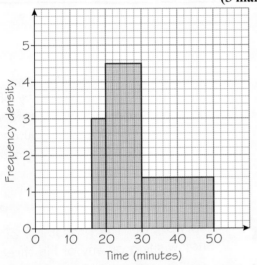

Histogram facts

☑ No gaps between the bars.

☑ **Area** of each bar is proportional to frequency.

☑ Vertical axis is labelled 'Frequency density'.

☑ Bars can be different widths.

☑ Frequency density = $\dfrac{\text{frequency}}{\text{class width}}$

The class widths in the frequency table are **different widths** so a histogram is the most suitable graph to draw.

1. Calculate the frequency densities.

$$\frac{12}{4} = 3 \qquad \frac{45}{10} = 4.5 \qquad \frac{28}{20} = 1.4$$

2. Add these values to the table as an extra column.

3. Label the vertical axis 'Frequency density'.

4. Choose a scale for the vertical axis and draw each bar.

Area and estimation

You can use the area under a histogram to estimate frequencies. An estimate for the **number** of maggots between 1 mm and 2 mm long is:

$$0.5 \times 22 + 0.5 \times 6 = 14$$

You might need to answer proportion questions about histograms in your exam. The total frequency is:

$$1 \times 14 + 0.5 \times 22 + 1.5 \times 6 = 34$$

So an estimate for the **proportion** of maggots between 1 mm and 2 mm long is: $\dfrac{14}{34} = 0.4117\ldots$ or 41% (2 s.f.)

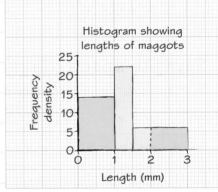

Now try this

Aiming higher

A speed camera recorded the speed of some vehicles on a road. The table on the right shows the results.

(a) On graph paper, draw a histogram to illustrate this data. **(3 marks)**

(b) Estimate the proportion of vehicles travelling at more than 65 km/h. **(2 marks)**

Speed, s (km/h)	Frequency
$0 < s \leqslant 40$	48
$40 < s \leqslant 50$	32
$50 < s \leqslant 70$	128
$70 < s \leqslant 80$	88
$80 < s \leqslant 110$	24

Probability

For **equally likely outcomes** the probability (P) that something will happen is:

$$\text{Probability} = \frac{\text{number of successful outcomes}}{\text{total number of possible outcomes}}$$

If you know the probability that an event **will** happen, you can calculate the probability that it won't happen:

P(Event doesn't happen) = 1 − P(Event happens)

 The probability that this spinner lands on 4 is $\frac{1}{4}$. So the probability that it **does not** land on 4 is $1 - \frac{1}{4} = \frac{3}{4}$

Add or multiply?

Events are **mutually exclusive** if they can't **both** happen at the same time. For mutually exclusive events:

P(A or B) = P(A) + P(B)

Events are **independent** if the outcome of one doesn't affect the outcome of the other. For independent events:

P(A and B) = P(A) × P(B)

Worked example

Amir designs a game for his school fair. This table shows the probability of winning a prize.

Prize	Badge	Keyring	Cuddly toy
Probability	0.35	0.18	0.07

(a) What is the probability of **not** winning a prize? **(2 marks)**

P(Win) = 0.35 + 0.18 + 0.07 = 0.6

P(Not win) = 1 − 0.6 = 0.4

(b) Ameena plays the game three times. What is the probability that she does not win a prize? **(2 marks)**

0.4 × 0.4 × 0.4 = 0.064

(a) To work out the probability of winning **any** prize you need to add together the probabilities. Then you can use this rule to work out the probability of not winning a prize:

$$P\left(\begin{array}{c}\text{Not winning}\\\text{a prize}\end{array}\right) = 1 - P\left(\begin{array}{c}\text{Winning a}\\\text{prize}\end{array}\right)$$

(b) To work out the probability of Amir not winning on any of his three games, you need to multiply the probabilities.

Sample space diagrams

A **sample space diagram** shows you all the possible outcomes of an event. Here are all the possible outcomes when two coins are flipped.

First coin

	H	T
H	HH	TH
T	HT	TT

Second coin

• There are four possible outcomes. TH means getting a tail on the first coin and a head on the second coin.

• The probability of getting two tails when two coins are flipped is $\frac{1}{4}$ or 0.25. There are four possible outcomes and only one successful outcome (TT).

Now try this

The table shows the probabilities of different delivery times for a letter.

Delivery time	Next day	1 day late	More than 1 day late
Probability	$45x$	$3x$	$2x$

Work out the probability of a letter arriving one day late. **(3 marks)**

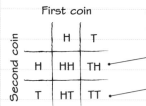

Don't just find x. You need to answer the question by writing the probability of the letter arriving one day late.

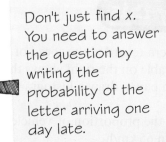 Worked solution video

Relative frequency

You need to be able to calculate probabilities for data given in graphs and tables. You can use this formula to estimate a probability from a frequency table:

$$\text{Probability} = \frac{\text{frequency of outcome}}{\text{total frequency}}$$

When a probability is calculated like this it is sometimes called a **relative frequency**.

Golden rule

Probability estimates based on relative frequency are **more accurate** for larger samples (or for a larger number of trials in an experiment).

In the sample there were 15 + 10 = 25 eggs which weighed 55 g or more. So an estimate for the probability of picking an egg that weighs 55 g or more is $\frac{25}{40}$ or $\frac{5}{8}$.

An egg farm weighed a sample of 40 eggs. It recorded the results in a frequency table.

Weight, w (g)	Frequency
$45 \leqslant w < 50$	6
$50 \leqslant w < 55$	9
$55 \leqslant w < 60$	15
$60 \leqslant w < 65$	10

(a) Roselle buys some eggs from the farm and picks one at random. Estimate the probability that the egg weighs 55 g or more. **(2 marks)**

$$P(w \geqslant 55) \approx \frac{25}{40} = \frac{5}{8}$$

(b) Comment on the accuracy of your estimate. **(1 mark)**

40 is a fairly small sample size, so the estimate is not very accurate.

Experimental probability

You can carry out an experiment to estimate the probability of something happening. This table shows the results of throwing a drawing pin 60 times.

Number of trials	10	20	30	40	50	60
Frequency of landing point up	8	11	17	25	30	37

To estimate the probability that the drawing pin will land point up, you calculate the relative frequency. The most accurate estimate will be based on the largest number of trials.

A four-sided spinner is spun 40 times.
The results are shown in the table.

Number	1	2	3	4
Frequency	8	4	21	7

(a) Work out the estimated probability of getting a 3. **(1 mark)**

(b) Work out the theoretical probability of getting a 3 on a fair four-sided spinner. **(1 mark)**

(c) Do you think this spinner is fair? Give a reason for your answer. **(1 mark)**

If the spinner was fair, then the experimental probability would get closer to the theoretical probability as the number of trials increased.

Set notation

In mathematics a **set** is a collection of **elements** (or **members**). The elements in a set could be numbers, words or letters. You can define a set in **two** different ways:

1 **Listing the elements**

A = {onions, carrots, peas} ── • You use curly brackets to define a set.

B = {13, 14, 15, 16} ── • Elements are separated by commas.

• This set contains 4 elements. You write n(B) = 4

2 **Using a rule**

C = {months with exactly 30 days} ──

D = {odd numbers between 10 and 20}

June is a member of this set.
You can write **June** ∈ C. The symbol ∈ means 'is a member of'. The symbol ∉ means 'is not a member of'.

• You could also write set D as {11, 13, 15, 17, 19}. It has 5 members so you write n(D) = 5

Union and intersection

U means **union**. The union of two sets is the set of elements that belong to **either** set.

∩ means **intersection**. The intersection of two sets is the set of elements that belong to **both** sets.

All or nothing

The symbol ℰ represents the **universal set**. This is the set of all the elements that you are allowed to consider in a particular question.

The symbol Ø is used to represent the **empty** (or **null**) set. It contains **no elements**.

Worked example

A = {r, o, m, e}
B = {p, r, a, g, u, e}

List the members of the set

(a) A ∩ B   **(1 mark)**

A ∩ B = {r, e}

(b) A ∪ B   **(1 mark)**

A ∪ B = {r, o, m, e, p, a, g, u}

Worked example

ℰ = {positive whole numbers **less than** 15}
X = {multiples of 5}
Y = {multiples of 3}

(a) Is it true that 20 ∈ X? Explain your answer.
                             (1 mark)

No. The universal set is numbers less than 15, so 20 cannot be a member of any set.

(b) Is it true that X ∩ Y = Ø? Explain your answer.
                             (1 mark)

Yes. X = {5, 10} and Y = {3, 6, 9, 12}. These sets have no common members, so their union is the empty set.

Problem solved!

Make sure you **answer the question** and give an **explanation**. Write 'yes' or 'no' then write an explanation in words that shows that you understand what the symbols used in the question mean.

> You will need to use problem-solving skills throughout your exam – **be prepared!**

Now try this

1 P = {m, e, t, r, i, c}
   Q = {i, g, c, s, e}

   List the members of the set
   (a) P ∩ Q   **(1 mark)**
   (b) P ∪ Q   **(1 mark)**

2 A = {factors of 20}   B = {prime numbers}
   (a) Is it true that A ∩ B = Ø?
       Explain your answer. **(1 mark)**
   (b) Work out n(A ∩ B)   **(1 mark)**

> P ∪ Q means elements that are either in P, or in Q, or in both.

> n(A ∩ B) means the number of elements in A ∩ B

Venn diagrams

You can represent two or more sets on a **Venn diagram**.
This Venn diagram shows two sets A and B:

The **universal** set ($\mathcal{E}$) is drawn as a rectangle around the other sets.

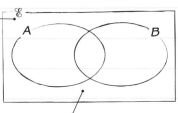

You draw each set as a circle. If the circles overlap then the sets **intersect**.

This shaded area represents $A \cap B$

This shaded area represents $A \cup B$
There is more on **intersection** and **union** on page 110.

Complements of sets

The complement of a set P is written as P′. You say 'not P' or 'the complement of P'.

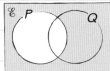

The shaded area represents P′.

Subsets

If X is a subset of Y it is **completely contained** inside Y. You write $X \subset Y$

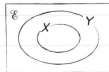

Every member of X is also a member of Y.

Worked example

A, B and C are sets, with $B \cap C = \varnothing$ and $A \subset B$

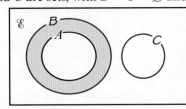

(a) Complete the Venn diagram to show sets B and C   **(2 marks)**

(b) On the Venn diagram, shade the region that represents $B \cap A'$ **(1 mark)**

Worked example

This Venn diagram shows three sets.

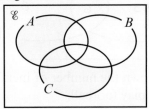

Use set notation to describe the shaded region.   **(2 marks)**

$(A \cup C) \cap B'$

This means 'A or C and not B'.

Now try this

On separate copies of this Venn diagram, shade the region represented by

(a) $P \cup Q$ **(1 mark)**

(b) Q' **(1 mark)**

(c) $R \cap P'$ **(1 mark)**

(d) $P \cup (Q \cap R)$ **(1 mark)**

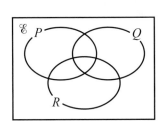

Probability and sets

You can write **numbers** in the sections of a Venn diagram to show the **number of elements** (or **members**) that are contained in that section. This Venn diagram shows whether a group of 50 people own a fish, a cat or both.

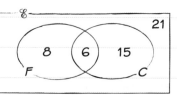

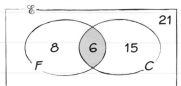

6 people owned a fish **and** a cat. You can write

$n(F \cap C) = 6$

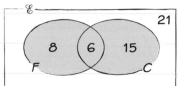

$8 + 6 + 15 = 29$ people in total owned a fish **or** a cat. You can write $n(F \cup C) = 29$

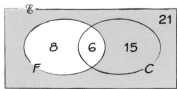

$15 + 21 = 36$ people **did not** own a fish. You can write $n(F') = 36$

Worked example

36 members of a youth club were surveyed about the sports they played.

19 members played tennis.

14 members played football.

6 members played both tennis and football.

(a) Draw a Venn diagram to show this information. **(3 marks)**

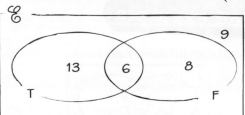

(b) Write down the number of members who did not play football. **(1 mark)**

$13 + 9 = 22$

(c) One of the members is chosen at random. Write down the probability that the member plays neither tennis nor football. **(1 mark)**

$\frac{9}{36} = \frac{1}{4}$

$T \cap F$ ────────────

Fill in the centre of the Venn diagram first. 6 members play both sports.

$T \cap F'$ ────────────

The 19 members who play tennis include the 6 members who play both sports. So $19 - 6 = 13$ members play tennis but *not* football.

$T' \cap F$ ────────────

$14 - 6 = 8$ members play football but *not* tennis.

$T' \cap F'$ ────────────

$36 - 13 - 6 - 8 = 9$ members play neither sport.

Check it!

$13 + 6 = 19$ members play tennis ✓

$8 + 6 = 14$ members play football ✓

$13 + 6 + 8 + 9 = 36$ members surveyed ✓

Now try this

The Venn diagram shows information about how 30 people in an office travel to work.

One person is chosen at random. Find the probability that this person

(a) uses both the bus and the train on their journey **(1 mark)**

(b) uses the bus at some point in their journey. **(2 marks)**

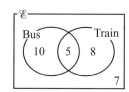

Conditional probability

If one event **has already occurred**, the probability of other events occurring might **change**. This is called conditional probability. The probability that an event X occurs **given that** an event Y has **already** occurred is written as $P(X|Y)$.

Using Venn diagrams

You can solve some conditional probability problems using a Venn diagram. If an event has already occurred, then the sample space for the other events is **shrunk**. These Venn diagrams show the outcomes of two events, A and B.

Complete sample space

$$P(A) = \frac{8+3}{8+3+4+5}$$

$$= \frac{11}{20}$$

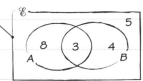

Event B occurs

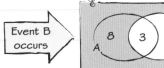

Shrunk sample space **given that** event B has occurred

$$P(A\,|\,B) = \frac{3}{3+4} = \frac{3}{7}$$

This Venn diagram shows the burger toppings chosen by a group of 50 diners at a restaurant. The choices are avocado (A), blue cheese (B) and chilli sauce (C).

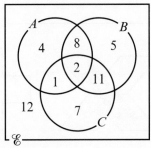

A diner is chosen at random.

(a) Given that the diner chooses blue cheese, find the probability that she also chooses avocado. **(2 marks)**

$$\frac{8+2}{8+2+11+5} = \frac{10}{26} = \frac{5}{13}$$

A second diner is chosen at random.

(b) Given that the diner chooses at least one of the three toppings, find the probability that she chooses all three. **(3 marks)**

$$\frac{2}{4+8+5+1+2+11+7} = \frac{2}{38} = \frac{1}{19}$$

(a) Look at this shrunk sample space:

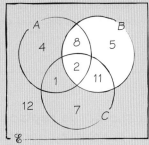

Of the $8+2+11+5 = 26$ diners who chose blue cheese, $8+2 = 10$ also chose avocado.

> You will need to use problem-solving skills throughout your exam – **be prepared!**

Two-way tables

You might have to work out probabilities from a two-way table. This table shows how a group of students travel to school.

	Car	Bus	Walk	Bike
Male	4	17	20	9
Female	8	20	29	11

Given that a student bikes to school, the probability that they are male is $\dfrac{9}{9+11} = \dfrac{9}{20}$

---

1 Look at the Venn diagram in the worked example above. A third diner is chosen at random. Given that the diner does **not** choose blue cheese, find the probability that she chooses chilli sauce. **(3 marks)**

2 Look at the two-way table in the blue box above. A student is chosen at random from this group. Given that the student is male, find the probability that he walks to school. **(3 marks)**

Tree diagrams

You can use a tree diagram to answer questions involving **conditional probability**.

A tree diagram shows all the possible outcomes from a series of events and their probabilities.

This is a tree diagram for Holly's journey to school.

You write the probability for each event on the branch.

At each branch the probabilities add up to 1.
$\frac{2}{3} + \frac{1}{3} = 1$

The outcome of the first event can affect the probability of the second. Holly is less likely to be on time if she misses the bus.

You write the outcomes at the ends of the branches. You can use shorthand like this.

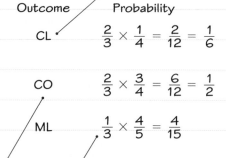

Catch bus
$\frac{2}{3}$
$\frac{1}{4}$ Late
$\frac{3}{4}$ On time

Miss bus
$\frac{1}{3}$
$\frac{4}{5}$ Late
$\frac{1}{5}$ On time

Outcome	Probability
CL	$\frac{2}{3} \times \frac{1}{4} = \frac{2}{12} = \frac{1}{6}$
CO	$\frac{2}{3} \times \frac{3}{4} = \frac{6}{12} = \frac{1}{2}$
ML	$\frac{1}{3} \times \frac{4}{5} = \frac{4}{15}$
MO	$\frac{1}{3} \times \frac{1}{5} = \frac{1}{15}$

Each branch is like a different parallel universe. In this universe, Holly catches the bus and gets to school on time.

You multiply along the branches to find the probability of each outcome.
The probability that Holly misses the bus and is late for school is $\frac{4}{15}$

Golden rules

1 Look out for the words **replace** or **put back** in a probability question.

With replacement: probabilities stay the same.

Without replacement: first probability stays the same while the others change.

2 ✕ MULTIPLY ALONG THE BRANCHES ✚ ADD UP THE OUTCOMES

Worked example

Aiming higher

There are 3 strawberry yoghurts and 4 pineapple yoghurts in a fridge. Bhaskor picks two yoghurts at random.
Work out the probability that both the yoghurts are the same flavour. **(4 marks)**

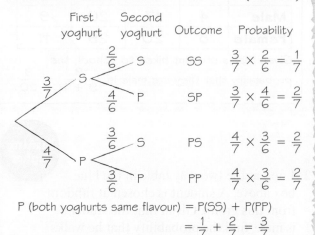

First yoghurt | Second yoghurt | Outcome | Probability
$\frac{3}{7}$ S
$\frac{2}{6}$ S → SS → $\frac{3}{7} \times \frac{2}{6} = \frac{1}{7}$
$\frac{4}{6}$ P → SP → $\frac{3}{7} \times \frac{4}{6} = \frac{2}{7}$
$\frac{4}{7}$ P
$\frac{3}{6}$ S → PS → $\frac{4}{7} \times \frac{3}{6} = \frac{2}{7}$
$\frac{3}{6}$ P → PP → $\frac{4}{7} \times \frac{3}{6} = \frac{2}{7}$

P (both yoghurts same flavour) = P(SS) + P(PP)
$= \frac{1}{7} + \frac{2}{7} = \frac{3}{7}$

This is an example of selection **without replacement**. The two events are not independent. The probabilities for the second pick change depending on which flavour yoghurt was picked first. A tree diagram is the **safest** way to answer questions like this.

Now try this

Worked solution video

Aiming higher

The probability that a student at Jen's school has a fish is 0.3.
If a student has a fish, the probability that they have a cat is 0.12.
If a student does not have a fish, the probability that they have a cat is 0.25.
A student is chosen at random.
Work out the probability that they do not have a cat. **(4 marks)**

Problem-solving practice 1

In your International GCSE Maths exams you will need to demonstrate **problem-solving** and **reasoning skills**. If you come across a tricky or unfamiliar question in your exam, you can try some of these strategies:

- ✓ Sketch a diagram to see what is going on.
- ✓ Try the problem with smaller or easier numbers.
- ✓ Plan your strategy before you start.
- ✓ Write down any formulae you might be able to use, or check the formulae sheet for a clue.
- ✓ Use x or n to represent an unknown value.

 Eight numbers have a mean of 12.
The numbers are

7   4   17   20 x 9   6 $2x$

Work out the value of x. **(4 marks)**

Mean, median and mode page 101

There are 8 values and the mean is 12. This means that the sum of the data values is $8 \times 12 = 96$

You know that all the values, including x and $2x$, must add up to 96. Use this information to write an equation and solve it to find x.

TOP TIP

You can solve lots of problems involving the mean using:

Sum of values = mean $\times$ number of values

 Huan spins this spinner. He keeps spinning until he lands on the unshaded sector.

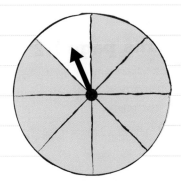

Work out the probability that Huan spins the spinner

(a) exactly twice   **(2 marks)**

(b) more than twice.   **(2 marks)**

Probability page 106

You might need to think carefully before working out which probabilities to multiply.

(a) Huan stops when he lands on white, so in order to spin the spinner **exactly** twice he must land on blue on the first spin, **and** white on the second spin.

(b) In order for Huan to spin more than twice he must land on blue on **both** of the first two spins.

TOP TIP

Use notation like P(Blue) and P(White) in probability questions to show your working clearly.

Problem-solving practice 2

 The weights in grams of the lemons in a box of lemons from Sunvale Farm are given in order below.

77 77 79 80 82 82 83 85
85 87 89 91 91 92 95

This table summarises the weights of the lemons in a box from Greentree Farm.

| Median | 84 grams |
| Interquartile range | 6 grams |

Compare the weights of the lemons in each box. **(5 marks)**

Interquartile range page 103

You need to calculate both of these statistics for the box from Sunvale Farm. There are 15 lemons in the box:

Lower quartile $= \dfrac{15 + 1}{4} = $ 4th value

Upper quartile $= \dfrac{3(15 + 1)}{4} = $ 12th value

TOP TIP

When comparing data make sure you calculate the **same** statistics for each data set, and support all your statements with **statistical evidence**.

 Beccy has two bags of numbered balls.

She chooses a ball at random from bag A and places it in bag B. Then she chooses a ball at random from bag B and places it in bag A.

Calculate the probability that the sum of the balls in bag A will be 18 or greater. **(4 marks)**

Aiming higher

Probability page 106

Work out all the ways Beccy can end up with a total of 18 or more in bag A.
Use a table to keep track of your working:

Work out the probability of each outcome and add them together.

A → B	B → A	Total in A
1	4	18
1	5	19
2	5	18

You can answer lots of tricky probability questions by writing down all the successful outcomes then working out the probability of each one.

---

⑤ In a board game, Diego picks a question card.
He can pick an easy (E), medium (M) or hard (H) question.

The probabilities of picking each type of question are 0.54, 0.31 and 0.15 respectively.

The probabilities that he gets each type of question correct are 0.8, 0.5 and 0.1 respectively.

Work out the probability that Diego gets his question correct. **(5 marks)**

Aiming higher

Tree diagrams page 112

Either draw a tree diagram or write down the successful outcomes and work out their probabilities. For example:

$P(E, \text{correct}) = 0.54 \times 0.8 = 0.432$

TOP TIP

If you draw a tree diagram for a conditional probability question you will probably get some or all of the marks for this question because you won't leave out any of the possibilities.

Formulae sheet

Arithmetic series
Sum to n terms, $S_n = \dfrac{n}{2}[2a + (n-1)d]$

Area of trapezium $= \dfrac{1}{2}(a + b)h$

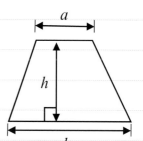

The quadratic equation
The solutions of $ax^2 + bx + c = 0$ where $a \neq 0$ are given by:

$$x = \frac{-b \pm \sqrt{(b^2 - 4ac)}}{2a}$$

Trigonometry

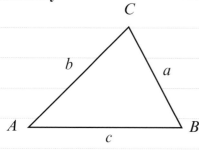

Volume of a prism
$= $ area of cross section $\times$ length

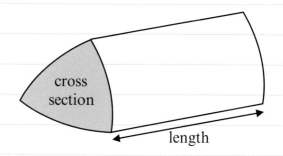

In any triangle ABC

Sine Rule $\quad \dfrac{a}{\sin A} = \dfrac{b}{\sin B} = \dfrac{c}{\sin C}$

Cosine Rule $\quad a^2 = b^2 + c^2 - 2bc \cos A$

Area of triangle $= \dfrac{1}{2}ab \sin C$

Volume of cone $= \dfrac{1}{3}\pi r^2 h$
Curved surface area of cone $= \pi r l$

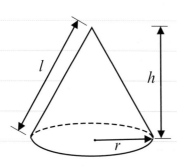

Volume of sphere $= \dfrac{4}{3}\pi r^3$

Surface area of sphere $= 4\pi r^2$

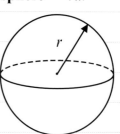

Volume of cylinder $= \pi r^2 h$

Curved surface area of cylinder $= 2\pi r h$

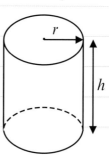

115

Answers

NUMBER

1. Factors and primes
1. (a) $2^2 \times 5 \times 7^2$ (b) 28
2. (a) $3^2 \times 7 = 63$ (b) $2 \times 3^5 \times 5 \times 7^2 = 119\,070$

2. Indices 1
1. (a) 7^8 (b) $n = 3$
2. $k = 2$
3. (a) (i) 2^4 (ii) 7^{12} (iii) 5^2
   (b) $n = 11$

3. Indices 2
1. (a) 7^{h-k} (b) 7^{2h} (c) 7^{h+2k}
2. $n = -3$
3. $\sqrt{\dfrac{49}{7^3}} = \sqrt{\dfrac{7^2}{7^3}} = \sqrt{7^{-1}} = 7^{-\frac{1}{2}}$

4. Calculator skills 1
(a) $4.197\,530\,864$ (b) 4.20 (3 s.f.)

5. Fractions
1. (a) $\dfrac{7}{10} - \dfrac{1}{4} = \dfrac{14}{20} - \dfrac{5}{20}$
   $= \dfrac{9}{20}$
   (b) $3\dfrac{4}{9} + 1\dfrac{5}{6} = \dfrac{31}{9} + \dfrac{11}{6}$
   $= \dfrac{62}{18} + \dfrac{33}{18}$
   $= \dfrac{95}{18}$
   $= 5\dfrac{5}{18}$
   (c) $\dfrac{3}{4} \div \dfrac{5}{12} = \dfrac{3}{4} \times \dfrac{12}{5}$
   $= \dfrac{9}{5}$
   $= 1\dfrac{4}{5}$
   (d) $1\dfrac{7}{8} \times 2\dfrac{2}{3} = \dfrac{15}{8} \times \dfrac{8}{3}$
   $= 5$
2. (a) $\dfrac{9}{20}$ (b) $120\,\text{cl}$

6. Decimals
1. e.g. $\dfrac{1}{250} = \dfrac{1}{2 \times 5^3}$. Only factors in denominator are 2 and 5 so terminating decimal.
2. e.g. $\dfrac{1}{140} = \dfrac{1}{2^2 \times 5 \times 7}$. Factors in denominator other than 2 and 5 so recurring decimal.
3. $11\,\overline{\smash{\big)}\,5.{}^5 0 {}^5 0 {}^5 0 {}^6 0 \cdots}$ giving $0.4\,5\,4\,5\ldots$
   $0.454\,545\ldots = 0.\dot{4}\dot{5}$

7. Recurring decimals
1. Let $n = 0.545\,454\,54\ldots$
   $100n = 54.545\,454\ldots$
   $99n = 54$
   $n = \dfrac{54}{99} = \dfrac{6}{11}$
2. Let $n = 0.018\,181\,81\ldots$
   $100n = 1.818\,181\,818\ldots$
   $99n = 1.8$
   $n = \dfrac{1.8}{99} = \dfrac{1}{55}$

3. Let $n = 0.351\,351\,351\ldots$
   $1000n = 351.351\,351\,351\ldots$
   $999n = 351$
   $n = \dfrac{351}{999} = \dfrac{13}{37}$

8. Surds 1
1. $\sqrt{32} + \sqrt{98} = \sqrt{16 \times 2} + \sqrt{49 \times 2}$
   $= 4\sqrt{2} + 7\sqrt{2}$
   $= 11\sqrt{2}$
   $p = 11$
2. $\dfrac{35}{\sqrt{7}} = \dfrac{35\sqrt{7}}{7} = 5\sqrt{7}$
3. $x = 32$

9. Calculator skills 2
1. 86.3%
2. Aisha: \$557.50; Javier: \$540; Aisha spends more on rent.

10. Ratio
1. £161
2. (a) 21
   (b) 24

11. Proportion
e.g. €16.55 = £14.39
1.5 lb = 0.681 81... kg
Cost in £ per kg in France = 14.39 ÷ 1.25 = 11.513
Cost in £ per kg in England = 8.97 ÷ 0.681... = 13.156
The cheese is cheaper in France

12. Percentage change
1. 32.1% (1 d.p.)
2. 59.6% (1 d.p.)

13. Reverse percentages
1. €45
2. €220 000

14. Repeated percentage change
(a) £5304.50
(b) 4 years

15. Upper and lower bounds
1. LB for width = LB for area ÷ UB for length
   $= 315 \div 22.5$
   $= 14\,\text{cm}$
2. $13.5\,\text{m}^2$

16. Standard form
(a) 277 000 (b) $5.511 \times 10^5\,\text{kg}$

17. Problem-solving practice 1

1. $96\,\text{mm}$
2. $2 \times \frac{3}{8} = \frac{3}{4}\,\text{kg per day.}$
   $14 \div \frac{3}{4} = 18\frac{2}{3}$. Susan can feed the cats for 18 days from 1 bag.
3. $n = 0.928\,282\,8\ldots$
   $100n = 92.828\,282\,8\ldots$
   $99n = 91.9$
   $n = \frac{91.9}{99} = \frac{919}{990}$

18. Problem-solving practice 2

4. 24.72 litres
5. 3.6 litres
6. $178\,\text{cm}$ (3 s.f.)

ALGEBRA

19. Algebraic expressions

1. h^{12}
2. (a) $16a^{20}b^4$ (b) $15x^7y^9$ (c) $3d^6g^5$
3. (a) $4p^5$ (b) $\frac{1}{4x^3y^{\frac{2}{3}}}$

20. Expanding brackets

1. (a) $x^2 + 4x - 5$
   (b) $p^2 - 12p + 36$
2. (a) $x^3 + 12x^2 + 27x$
   (b) $n^3 + 11n^2 + 39n + 45$

21. Factorising

1. (a) $(x + 10)(x + 2)$ (b) $(x + 2)(x - 5)$
2. (a) $3g(4 + g)$ (b) $(p - 14)(p - 1)$ (c) $2x(3x - 4y)$
3. (a) $4ma(1 - 6m)$ (b) $(p + 8)(p - 8)$
4. $(3x - 2)(x - 2)$

22. Linear equations 1

1. (a) $w = 7$ (b) $x = -1$
2. (a) $y = -\frac{7}{4}$ (b) $m = 4$

23. Linear equations 2

1. (a) $w = -5$ (b) $x = 3$
2. (a) $y = 6$ (b) $m = 37$

24. Formulae

$21.9\,\text{cm}$

25. Arithmetic sequences

(a) nth term $= 4n - 1$
(b) No. 22nd term $= 87$, 23rd term $= 91$
   (or $4n - 1 = 89$
          $n = 22.5$ which is not a whole number)

26. Straight-line graphs 1

$y = -2x + 2$

27. Straight-line graphs 2

1. $y = 6x - 20$
2. $y = 2x - 1$
3. $k = 6$

28. Parallel and perpendicular

1. $y = \frac{1}{2}x + 4$
2. (a) $y = \frac{4}{3}x$
   (b) $y = -\frac{3}{4}x + 12.5$

29. Quadratic graphs

(a)

x	-3	-2	-1	0	1	2	3
y	11	6	3	2	3	6	11

(b)

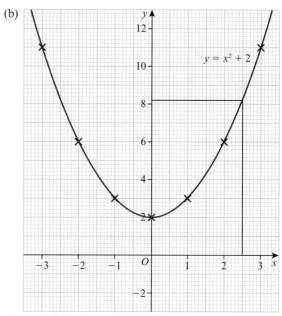

(c) $y = 8.2$ (to the nearest small square)

30. Cubic and reciprocal graphs

1. (a) D   (b) E   (c) A   (d) B   (e) F   (f) C

31. Rates of change

(a) 30 minutes
(b) $30\,\text{km/h}$

32. Quadratic equations

(a) $m = 6, m = 2$ (b) $w = -4, w = 9$
(c) $y = \frac{3}{5}, y = -8$ (d) $x = \frac{5}{7}, x = -1$

33. The quadratic formula

1. (a) $x = -1.16, x = 0.74$ (2 d.p.)
   (b) $m = -106.39, m = 56.39$ (2 d.p.)
2. $x = 6.42, x = 0.0779$ (3 s.f.)

34. Completing the square

1. (a) $x = -5 \pm \sqrt{13}$
   (b) $y = 3 \pm 2\sqrt{6}$ (or $y = 3 \pm \sqrt{24}$)
2. $2(x + 5)^2 - 43$

35. Simultaneous equations 1

1. $x = 5$, $y = 1.5$

2.

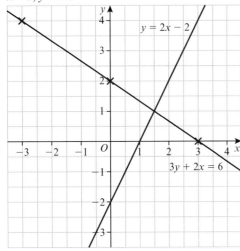

$x = 1.5$, $y = 1$

36. Simultaneous equations 2

1. $x = 5$, $y = 1$ and $x = -3$, $y = -3$
2. $(2, 3)$ and $\left(-\frac{4}{3}, \frac{7}{9}\right)$

37. Inequalities

1. (a) $n > -2$ (b) $n \geqslant 22$
2. $x = 8$
3. $n = -2, -1, 0, 1$
4. $x < \frac{1}{8}$

38. Trigonometric graphs

(a)

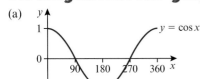

(b) 315°

39. Transforming graphs 1

(a) (i) $(0, -1)$
    (ii) $(2, -3)$
    (iii) $(1, -1)$
(b) $y = f(-x)$

40. Transforming graphs 2

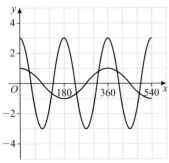

41. Inequalities on graphs

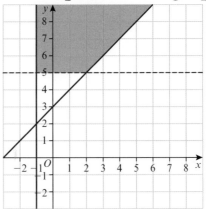

42. Sketching graphs

1. $-2 < p < 2$

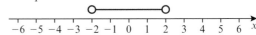

2. $-2 < x < 7$

43. Using quadratic graphs

(a)

x	-5	-4	-3	-2	-1	0	1	2
y	7	1	-3	-5	-5	-3	1	7

(b)

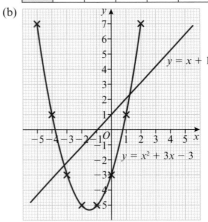

(c) $x = -3.8$, $x = 0.8$ (d) $x = -3.2$, $x = 1.2$

44. Turning points

(a) $(0, -3)$ (b) $(2, -7)$

45. Quadratic inequalities

(a)   (b)

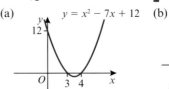

46. Gradients of curves

(a) 0.8 cm/s (accept 0.7–0.9)   (b) 0.5 cm/s (accept 0.4–0.6)

47. Proportion and graphs

(a) 4 inches   (b) 32 cm
(c) Graph is a straight line and passes through the origin $(0, 0)$

48. Proportionality formulae

1. (a) $x = 7.2y$ (b) 230.4   (c) 1.25
2. 8 cm

49. Harder relationships

1. (a) $t = 0.92\sqrt{d}$
   (b) 5.98 seconds
2. 3.09 km/s

50. Rearranging formulae

1. $t = \dfrac{4p + 1}{3}$

2. $w = \dfrac{m^2 - 7}{5}$

3. $P = \dfrac{100}{Q^2 + 5}$

51. Sequences and series

1. 4020
2. 11175

52. Algebraic fractions

1. $\dfrac{2a + 9}{6a}$
2. $x = \frac{1}{120}$
3. (a) $\dfrac{x(x - 2)}{x + 5}$ (b) $\dfrac{m - 6}{3m^2}$

53. Quadratics and fractions

1. $x = -\frac{5}{3}, x = 2$
2. $x = \frac{3}{4}, x = -3$
3. $x = -\frac{15}{2}, x = -4$

54. Surds 2

1. $(3 - \sqrt{12})^2 = (3 - \sqrt{12})(3 - \sqrt{12})$
   $= 9 - 3\sqrt{12} - 3\sqrt{12} + (\sqrt{12})^2$
   $= 21 - 6\sqrt{12}$
   $= 21 - 6 \times 2\sqrt{3}$
   $= 21 - 12\sqrt{3}$
2. $x = 3$ and $k = 8$

55. Functions

1. (a) $f(1) = 4$ (b) $x = 0$ (c) $b = 1.5$
2. (a) $f(3) = 16$ (b) $f(x) \geqslant 9$ (c) $x < 5$

56. Composite functions

1. (a) $gf(x) = 4x^2 - 8x$ (b) $x = 0, x = 2$
2. (a) $gf(x) = \dfrac{x + 1}{2x + 1}$ (b) $fg(x) = \dfrac{3x - 1}{2x - 1}$

57. Inverse functions

1. (a) $f^{-1}(x) = \dfrac{x + 1}{2}$ (b) 7
2. $g^{-1}: x \rightarrow = \dfrac{6}{x - 1}$

58. Differentiation

1. (a) $8x + 6$ (b) $-2x^{-2}$
2. (a) $x^{-2} + x^{-1}$ (b) $-2x^{-3} - x^{-2}$

59. Gradients and calculus

1. (a) $\dfrac{dy}{dx} = 9x^2 + 4x$ (b) 44
2. (0.8, 7.6)
3. (a) $\dfrac{dy}{dx} = 6x^2 - 6$
   (b) $(1, -3)$ and $(-1, 5)$

60. Turning points and calculus

1. (a) $(4, -13)$
   (b) Minimum, because the coefficient of x^2 is positive
2. (a) $\dfrac{dy}{dx} = 3x^2 - 10x + 8$
   (b) $x = 2, x = \frac{4}{3}$

61. Kinematics

(a) $\dfrac{ds}{dt} = 15 - 9.8t$ (b) 10.1 m/s   (c) 11.5 m (1 d.p.)

62. Algebraic proof

1. e.g. $\frac{1}{2}n(n + 1) + \frac{1}{2}(n + 1)((n + 1) + 1) = \frac{1}{2}n(n + 1) + \frac{1}{2}(n + 1)(n + 2)$
   $= \frac{1}{2}(n + 1)(n + (n + 2))$
   $= \frac{1}{2}(n + 1)(2n + 2)$
   $= \frac{1}{2}(n + 1) \times 2(n + 1)$
   $= (n + 1)(n + 1)$
   $= (n + 1)^2$
2. e.g. $(n + 4)^2 - n^2 = n^2 + 8n + 16 - n^2$
   $= 8n + 16$
   $= 8(n + 2)$
   The mean of $(n + 4)$ and $n =$ is $\dfrac{n + 4 + n}{2} = n + 2$
   Therefore, the difference between the squares, $8(n + 2)$, is 8 times the mean, $(n + 2)$

63. Problem-solving practice 1

1. The nth term formula is $6n + 3$
   Setting $6n + 3 = 65$ and solving for n gives $n = 10\frac{1}{3}$
   This is not an integer, so 65 is not a term in the sequence.
2. Time to travel y km = 3 minutes = 0.05 hours
   Speed $= \dfrac{\text{distance}}{\text{time}} = \dfrac{y}{0.05} = 20y$ km/h
3. (a) $y = 8x - 30$
   (b) (10, 50) does lie on the line.

64. Problem-solving practice 2

4. $(7a - 2b)(5a + 4b)$
5. (a) $\dfrac{x^2 - 2x}{3x^2 - 5x - 2} = \dfrac{x(x - 2)}{(3x + 1)(x - 2)}$
   $= \dfrac{x}{3x + 1}$
   $k = 3$
   (b) $f^{-1}(x) = \dfrac{x}{1 - 3x}$
6. (a) (i) $2x + 3y = 60$
   $3y = 60 - 2x$
   $y = 20 - \frac{2}{3}x$
   (ii) $A = xy$
   $= x(20 - \frac{2}{3}x)$
   $= 20x - \frac{2}{3}x^2$
   (b) $\dfrac{dA}{dx} = 20 - \frac{4}{3}x$
   (c) 150

GEOMETRY & MEASURES

65. Speed
86.4 km/h

66. Density
9.234 g/cm^3

67. Other compound measures
1. 16.6 km/l
2. 2.4 minutes or 2 minutes and 24 seconds

68. Angle properties
$x = 59°$

69. Solving angle problems
$\angle BDC = x$ (Base angles of an isosceles triangle are equal)
$\angle DBC = 180° - 2x$ (Angles in a triangle add up to 180°)
$\angle ABC = x$ (Base angles of an isosceles triangle are equal)
$\angle ABD = x - (180° - 2x) = 3x - 180°$

70. Angles in polygons
(a) $n = 36$　　　　　(b) 1440°

71. Perimeter and area
180 cm^2

72. Units of area and volume
1. (a) 23 000 cm^2　　　(b) 0.4 cm^3
2. 3.5×10^8 mm^3
3. 37 500 N/m^2

73. Prisms
(a) 600 cm^3　　　(b) 660 cm^2

74. Circles and cylinders
119 cm^2

75. Sectors of circles
12.6 cm (3 s.f.)

76. Pythagoras' theorem
(a) $y = 3.5$ cm　　　(b) $z = 9.1$ cm

77. Trigonometry 1
(a) 44.8°　　　(b) 56.2°　　　(c) 48.6°

78. Trigonometry 2
(a) 4.4 cm　　　(b) 5.4 cm　　　(c) 15.7 cm

79. Volumes of 3D shapes
806 mm^3 (3 s.f.)

80. Surface area
1. 878 cm^2
2. 115 cm^2

81. Translations, reflections and rotations
(a) Rotation, 90° clockwise, centre (2, 2)
(b) Translation with vector $\begin{pmatrix} -5 \\ -2 \end{pmatrix}$

82. Enlargements
(a), (b)

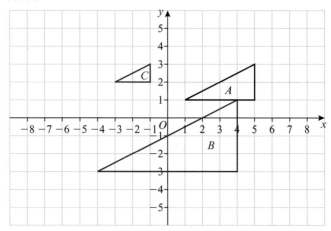

83. Combining transformations
Rotation 180° about point (4, 5)

84. Bearings
(a) 248°　　　(b) 079°

85. Scale drawings and maps
1. 4.08 km^2 (3 s.f.)
2. 43 km (accept 42–44 km)

86. Constructions

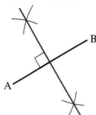

87. Similar shapes 1
(a) 130°　　　　　(b) 31.5 cm

88. Similar shapes 2
$W = 60$ cm　　　$X = 765$ cm^2　　　$Y = 290$ cm^3　　　$Z = 36 250$ cm^3

89. The sine rule
(a) 45.2° (3 s.f.)　　　(b) 15.2 cm (3 s.f.)

90. The cosine rule

Angle $PQR = 39.8°$ (3 s.f.)

91. Triangles and segments

Area of sector $= \frac{135}{360} \times \pi \times 4^2 = 6\pi$

Area of triangle $= \frac{1}{2} \times 4 \times 4 \times \sin 135°$

$\qquad = \frac{1}{2} \times 4 \times 4 \times \sin 45°$

$\qquad = \frac{1}{2} \times 4 \times 4 \times \frac{1}{\sqrt{2}}$

$\qquad = 4\sqrt{2}$

Area of segment $= 6\pi - 4\sqrt{2}$

92. Pythagoras in 3D

$DF = \sqrt{20^2 + 25^2 + 10^2} = \sqrt{1125} = 33.541\ldots$ cm > 33 cm

So the needle will fit inside the box.

93. Trigonometry in 3D

$27.1°$ (3 s.f.)

94. Circle facts

$63.5°$

95. Intersecting chords

(a) $x = 3.5$ (b) $y = 3.9$ (1 d.p.)

96. Circle theorems

angle BAC = angle $CBQ = x$ (Alternate segment theorem)

angle $BAD = 180° - 116° = 64°$ (Opposite angles of cyclic

$x + x + 8 = 64$ quadrilateral add up to 180°)

$\qquad x = 28°$

97. Vectors

(a) (i) $\begin{pmatrix} 6 \\ 10 \end{pmatrix}$

(ii) $\begin{pmatrix} 5 \\ -4 \end{pmatrix}$

(iii) $\begin{pmatrix} -7 \\ -24 \end{pmatrix}$

(b) 5.8

98. Vector proof

(a) $\overrightarrow{AB} = -2\mathbf{a} + 4\mathbf{b}$

$\overrightarrow{HA} = -\mathbf{b} + 2\mathbf{a}$ or $2\mathbf{a} - \mathbf{b}$

$\overrightarrow{JS} = 2\mathbf{b} - \frac{2}{3}(-2\mathbf{a} + 4\mathbf{b})$

$\qquad = 2\mathbf{b} + \frac{4\mathbf{a}}{3} - \frac{8\mathbf{b}}{3} = \frac{4\mathbf{a}}{3} - \frac{2\mathbf{b}}{3} = \frac{2}{3}(2\mathbf{a} - \mathbf{b})$

$\overrightarrow{KT} = \mathbf{b} - \frac{1}{3}(-2\mathbf{a} + 4\mathbf{b})$

$\qquad = \mathbf{b} + \frac{2\mathbf{a}}{3} - \frac{4\mathbf{b}}{3} = \frac{2\mathbf{a}}{3} - \frac{\mathbf{b}}{3} = \frac{1}{3}(2\mathbf{a} - \mathbf{b})$

$\overrightarrow{HA}$, $\overrightarrow{JS}$ and $\overrightarrow{KT}$ are all multiples of $(2\mathbf{a} - \mathbf{b})$, so are parallel.

(b) $KT:JS:HA = 1:2:3$

99. Problem-solving practice 1

1. 5.81 m (3 s.f.)
2. 37.5 cm^3

100. Problem-solving practice 2

3. Interior angle of pentagon $= 108°$
   Interior angle of A $= 360° - 108° - 90° = 162°$
   Exterior angle of A $= 180° - 162° = 18°$
   $n = 360° \div 18° = 20$

4. 31.7 cm^2 (3 s.f.)

5. $\angle AOB = 150°$ (Angle at the centre of a circle is twice the angle at the circumference)

   $\angle PBO = \angle PAO = 90°$ (Angle between a tangent and a radius is 90°)

   $\angle APB = 360° - 150° - 90° - 90°$ (Angles in a quadrilateral add up to 360°)

   $\qquad = 30°$

PROBABILITY & STATISTICS

101. Mean, median and mode

19

102. Frequency table averages

0.225 seconds

103. Interquartile range

(a) 19   (b) 5.4 cm   (c) 1.7 cm

104. Cumulative frequency

Median = 0.225 seconds   IQR = 0.045 seconds

105. Histograms

(a)

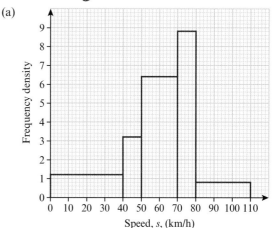

(b) $\dfrac{\left(\frac{128}{4} + 88 + 24\right)}{320} \times 100 = \dfrac{144}{320} \times 100 = 45\%$

106. Probability

0.06 or $\frac{3}{50}$

107. Relative frequency

(a) 0.525 or $\frac{21}{40}$
(b) 0.25 or $\frac{1}{4}$
(c) No. The experimental probability is much greater than the theoretical probability.

108. Set notation

1 (a) $P \cap Q = \{e, i, c\}$ (b) $P \cup Q = \{m, e, t, r, i, c, g, s\}$

2 (a) No. $A \cap B = \{2, 5\}$ (b) $n(A \cap B) = 2$

109. Venn diagrams

(a)
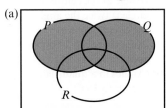

(c)
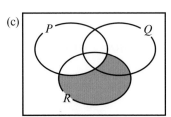

(b)
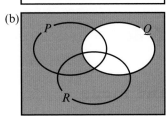

(d)
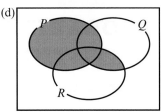

110. Probability and sets

(a) $\frac{5}{30} = \frac{1}{6}$

(b) $\frac{15}{30} = \frac{1}{2}$

111. Conditional probability

1. $\dfrac{1+7}{4+1+7+12} = \dfrac{1}{3}$

2. $\dfrac{20}{4+17+20+9} = \dfrac{2}{5}$

112. Tree diagrams

0.789

113. Problem-solving practice 1

1. $x = 11$

2. (a) $\frac{7}{64}$ (b) $\frac{49}{64}$

114. Problem-solving practice 2

3. Sunvale Farm median = 85 g
Sunvale Farm interquartile range = 11 g
The weights of the lemons from Sunvale farm were greater on average (median of 85 g vs 84 g) but more spread out (interquartile range of 11 g vs 6 g).

4. $\frac{3}{30} = \frac{1}{10}$

5. P(Correct) = $0.54 \times 0.8 + 0.31 \times 0.5 + 0.15 \times 0.1 = 0.602$

Notes

Notes

Notes

Notes